AMY JOHNSON:

HESSLE ROAD TOMBOY –

Born and Bred, Dread and Fled

[B&W: Black & White Version]

by ALEC GILL

Text copyright © 2016 Alec Gill
All Rights Reserved
No part of this book may be reproduced,
stored in a retrieval system or transmitted in any form,
or by any means electronic, mechanical, photocopying, recording
or otherwise without the prior permission of the Copyright holder.
Image copyrights are stated with each picture – where known.
If I have made any errors or omissions, then please let me know
so that the correct copyright status can be amended.
This publication forms part of my book series called:
HESSLE ROAD: Stories about Hull's Fishing Community and
Arctic Trawling Heritage (England)

WORDSPIN Productions, Hull - 2016

ISBN-13: 978-1535109949
ISBN-10: 1535109947

CONTENTS

	INTRODUCTION: "A Most Fertile Liar"	5
1	FISHING ANCESTORS: Danish Blood	12
2	FISH EMPIRE: Smelly Dock	16
3	BORN: A Tomboy in an Ugly Street	20
4	BRED: A Methodist Family	24
5	FORMATIVE YEARS: A Boulevard Scholar	29
6	FLED: Have Bike, Will Escape	35
7	DREAD: Cod Farm Lasses	39
8	ROOTLESS and RESTLESS: Alone in London	46
9	JOHNNIE: One of the Lads	53
10	LUCKY JASON: Have Plane, Will Escape	59
11	AUSTRALIA: Fingers-crossed Landings	65
12	FAMOUS and AIMLESS: Unlucky in Love	74
13	WAR and DEATH: Cut the Engines!	86
14	FUTURE: May Her Fame Live On	99
15	APPENDICES: Introduction	103
15A	Myself When Young by Amy Johnson	105
15B	At Amy's First School - interview with a Fellow Pupil	127
15C	New Year's Day in China by Amy Johnson	130
15D	Sky Roads of the World by Amy Johnson	134
15E	A Day's Work in the A.T.A. by Amy Johnson	154

15F	Letter to Women's Engineering Society by Amy Johnson	159
15G	Obituary Letter to The Times by Caroline Haslett	161
15H	Obituary Letter to The Times by Pauline Gower	163
15I	Anonymous Obituary by an ATA Pilot	165
15J	Extracts from Speeches by Amy Johnson	168
16	ACKNOWLEDGEMENTS: With Thanks	171
17	BIBLIOGRAPHY: My Own C-T-S Referencing Style	172
18	DEDICATION: Appreciation	176
19	ABOUT: The Author and Book	178
20	ENDNOTES: Sources and Snippets	181

HESSLE ROAD

INTRODUCTION: "A Most Fertile Liar"

Amy Johnson is world famous. "AMY, WONDERFUL AMY" was one of ten songs written about her[1.] Her rapid rise to stardom stemmed from her courageous solo flight from England to Australia in May 1930 in an open cockpit bi-plane – the first woman to do so. She became a global celebrity – and rightly so.

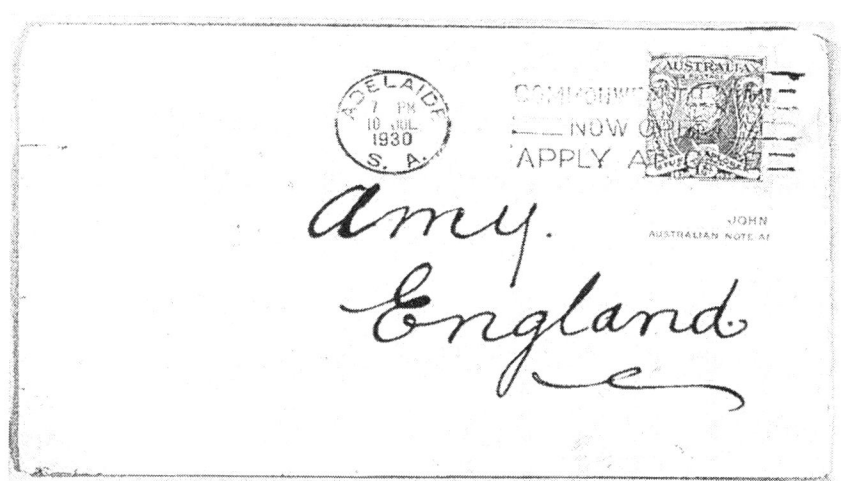

Image 1: Here is one tiny example of just how well known Amy was in her time. This is a fan letter dated 10 July 1930 that arrived all the way from Adelaide, Australia simply addressed to 'Amy, England'. Courtesy Hull Daily Mail.

Many books focus upon her aeronautical adventures and romantic dramas. I do not intend to duplicate Amy's life when she was in the public eye - except for some key elements of her exciting flight to Australia and her mysterious death. I would urge readers intrigued by her enigmatic activities to read some of the many biographies.

Most biographers take the stance that Amy was always moving towards two primary goals: her love of flying and her eagerness to break flight records to Australia and elsewhere. I take a different perspective. My belief is that Amy was constantly on the run from her Hull roots. Generally, biographers have avoided any in-depth analysis of Amy's girlhood; neither have they probed too much into her fishing family origins. Naturally, her global flying feats were what interested authors – not her back-story.

There is a Jesuit maxim that goes something like "Give me the child until seven years old and I will give you the adult". Amy Johnson's first seven, formative years (and far beyond) were all spent growing up in Hull's famous Hessle Road Fishing Community. Those vital, early years had a greater impact upon her temperament and achievements than has been previously recognised. It is my purpose to explore and outline the dynamic impact that the port's trawling culture had upon her character and determination. I am eager to expose Amy's secretive Hessle Road experience - one she tried to hide - and to link those times with her personality traits in adult life.

Our personality traits, I would suggest, are a bit like a massive jigsaw puzzle – made up of many segments. One difficulty is that even we ourselves rarely have the bigger picture of who we are. Even more difficult, as a biographer, is trying to piece together someone else's personality – especially when so many fragments are missing, we have never met that person, and must rely only on written text and hearsay.

Amy Johnson, like all of us, was a complex person. Labels drawn from her biographies include: courageous, restless, selfish, brave, foolhardy, determined, friendly, greedy, impetuous, flighty, apolitical, obstinate, funny, rebellious, patriotic, anarchistic, unworldly, and adventurous. The list goes on, but one trait Amy highlighted about herself was as "a most fertile liar"[2].

As a historian with a strict focus upon Hull's Fishing Community, I must include Amy in my work. For those who do not know, my dedicated research into the fishing culture has taken a variety of forms since I began over forty years ago. Initially, I was a social photographer (1974+) and have had twenty solo exhibitions based upon this documentary work; then author and historian (1982+); and filmmaker (1995+). In the early 1990s, I also became a wall-plate artist / designer - odd though that might sound. In 1992, I dedicated my second fine bone china plate (made at a Stoke-on-Trent pottery[3]) to "The Women of Hessle Road" – it comprised eight sketches (Image 2).

Image 2: The Women of Hessle Road wall plate has four (colourful) sketches in the centre that all focused upon life in the community: a midwife with her bicycle; two net braiders; a mother pushing a pram to the fish dock to collect her husband's weekly wage; and a herring girl splitting fish. Either side were two powerful female figures: (1) Big Lil Bilocca (right) campaigned for trawler safety after the 1968 triple tragedy, plus a drawing of the lost trawler St.Romanus (H.223); and (2) Amy Johnson the world-famous pilot, plus her aircraft Jason. Copyright Alec Gill.

Soon after the plate went on sale, a surprising question arose: "What the heck has Amy Johnson got to do with Hessle Road?" Hessle Roaders in particular queried her inclusion. They were baffled that glamorous Amy was born and bred in their community. Many people were aghast and implied that a serious historical faux pas had been committed on my part. What was going on? Did I have it wrong? Was it a serious mistake to place Amy on a wall plate as if she was a woman of Hessle Road?

Equally, was something else going on? Had Amy herself been intentionally misleading about her origins? Had she set up a massive smokescreen to blur her family roots? Had she manipulated her image to deceive the public? Was she embarrassed about her background and, therefore, fled from it? The answer is Yes. She certainly had something that caused her dread within her family. Fertile liar Amy succeeded in distorting history - and continues to do so well over 75 years after her death. As a researcher, I felt there was a mystery to solve. I needed to dig deeper and establish why Amy was deliberately deceptive about her past.

She did such a good job that many people today believe she was born and bred in the leafy Avenues area of Hull; others assume she grew up in Beverley. "In later life Amy was known to tell people that she originated from Bridlington, rather than admit she came from Hull"[4]. The very first national newspaper article about Amy's proposed flight to Australia reported that she came from "a wealthy Midlands family"[5]. During her solo flight, her parents suddenly became the focus of world media attention. They happened to live in Hull's Park Avenue district – a rich part of the city. People naturally assumed that Amy spent her childhood days there too. She never once corrected any such mis-information about her family background. Why? What was she trying to hide? Was she ashamed of her place of birth?

Because of Amy's lies, there is now a gross public mis-perception about her earliest days. She was, however, very much a Hessle Road girl – born and bred (1st July 1903). Her Victorian ancestors in the Johnson Family line helped lay the foundations of Hull as a trawling port. Indeed, the wealth of the family drew largely from the labour of the Hessle Road fishing families. Without good profits from her family's fish dock business, Amy would never have become world famous. To answer why she lied is not simple – and the truth will never be known. Nevertheless, I have investigated her life in various chapters – a synopsis of each one is listed below:

(1) FISHING ANCESTORS: Danish Blood - Amy was the third generation of a Danish immigrant. Her fishing family genealogy goes back to Denmark and her paternal grandparents (plus uncles, aunties and cousins) resided mainly in the Hessle Road area of Hull – as did she herself, of course.

(2) FISH EMPIRE: Smelly Dock - The Johnson's fish trade business dated back to Victorian England and the days of sailing smacks. Indeed, the business name Andrew Johnson Knudtzon (AJK) still exists in Hull over 135 years after its formation in 1881[6]. The firm has diversified many times and now has enormous cold storage distribution depots around the Hessle Road area in the Boulevard, Gillett Street and Neptune Street (near the busy A63 Clive Sullivan Way – Image 3).

(3) BORN: A Tomboy in an Ugly Street – Amy's birthplace (and two childhood homes) was in the heart of the Hessle Road Fishing Community – not too far from St.Andrew's Fish Dock (source of the family's prosperity). Amy's father Will Johnson, unwittingly, encouraged his first-born daughter to become a tomboy – like father, like daughter.

He pointed her towards unladylike activities. Amy had larger-than-life male role models from her dynamic Danish granddad and Klondyke-trekking father.

(4) BRED: A Methodist Family - The Johnsons had strong links with the Wesleyan Methodists. Amy's parents, Will and Ciss, were married at a Methodist Chapel on Hessle Road. They were both active members of the congregation. Amy and her sisters attended the chapel's Sunday School.

(5) FORMATIVE YEARS: A Boulevard Scholar - After attending some private schools, Amy was formerly educated at the Boulevard School (until she was a young woman of 19 years). She mixed with her peers from the fishing families and was attracted to the local lads. Initially, school was fun and she was the centre of attention. She was very sporty and became involved in many team games. Suddenly, it all changed and Amy became glum and isolated from her former classmates.

(6) FLED: Have Bike, Will Escape – Amy's school experience of being socially ridiculed (dread), led her to react by adopting a mechanical means of escaping from embarrassing situations (fled). She did not stay and solve her problems, she ran away from them. That was her primary 'coping mechanism'. Escape was how she reacted to a humiliating crisis. Amy's community childhood and Boulevard schooldays laid firm foundations for the way she coped in her adult life. She also drifted away from her Methodist upbringing and absorbed a variety of superstitious beliefs associated with fishing folklore.

(7) DREAD: Cod Farm Lasses - The wealth of the Johnson Family came primarily from trading in smelly fish. AJK's Cod Farm covered a vast fish dock estate and employed scores of workers. Amy's mother Ciss came from a genteel social background and cushioned her four daughters from contact with the fish dock and any links with the notorious Cod Farm Lasses. Amy loathed her family's fish business. She never really acknowledged AJK's crucial, financial support for her Australian venture.

(8) ROOTLESS and RESTLESS: Alone in London – After leaving school (aged 19) Amy was adrift. She had no real idea what to do with her life – or where to go. Her undergraduate days at Sheffield University were not too successful. After returning to Hull, mundane office work drove her 'around the bend'. London was merely another escape route to nowhere. Amy bounced from job to job, from digs to digs. Family friction and her two-timing, long-term, foreign lover made matters worse.

(9) JOHNNIE: One of the Lads – Some stability entered Amy's life soon after she found the Stag Lane Aerodrome in North London, especially its engineering workshops. At last, she had purpose and direction. After overcoming a mountain of macho attitudes against her, tomboy Amy was nicknamed Johnnie. She became the first woman in the world to qualify as an engineer in aircraft maintenance. Yet some in her

close family and circle of friends feared that her flying ambitions were a disguised suicide attempt to end all her problems.

(10) LUCKY JASON: Have Plane, Will Escape - Amy's aircraft Jason had Hull's fishing industry stamped all over it – though few (to this day) understand its secret fish dock significance. The world media were quick to link Jason and the Argonauts with Amy's quest to reach 'The Land of the Golden Fleece' – Australia with all its sheep! They were wrong.

(11) AUSTRALIA: Fingers-crossed Landings – Amy's Hessle Road determination and survival instincts kept her going forward. Amy and Jason battled against horrendous obstacles. She defied the gods and risked the odds. She landed in Australia on Empire Day to a tremendous welcome (24 May 1930).

(12) FAMOUS and AIMLESS: Unlucky in Love – Amy was now a global star and life was never the same again. She mixed with upper-class society, but repeated old personality traits. She failed in love and marriage. Amy was never true to herself. She shunned her parents who had supported her the most when she was in poverty and isolated. She turned her back on Hull and was ashamed of her Yorkshire accent. She could flee others, but not herself.

(13) WAR and DEATH: Cut the Engines! – The Second World War gave Amy direction, purpose and happiness – sadly, it was short-lived. Amy foretold how she would die. Even at the moment of her mysterious death in the Thames Estuary, a Hull trawlerman tried to save her life. This is an untold story that officials did not want to hear.

(14) FUTURE: May Her Fame Live On – A variety of colourful projects will ensure that Amy's fame will live on for generations to come. There are two female pilots aiming to replicate Amy's flight to Australia; a festival dedicated to Amy during 2016, and an attempt to find the wreck of her World War Two aircraft.

(15) APPENDICES (A to J): Amy's Own Writing and Other Authors – Ten appendices provide examples of Amy as an author (from childhood to when she served in the war) plus obituaries after her unexpected death. These samples are from rarely seen publications and enable readers to double check my source material.

Amy Johnson was a Hessle Roader – whether she liked it or not. From a time decades before her birth, her fishing family's prosperity, her schooling, large chunks of her short life, her Australian flight, her superstitious outlook, and at the instant of her death, all had links to Hessle Road. We cannot deny our roots. She tried, she lied; but perhaps that was why she never found true love or happiness. My plan is to re-attach Amy to her real Hessle Road heritage.

Image 3: There is some gratification (for me) in knowing that the firm founded by Amy Johnson's Danish grandfather in 1881 is still in business today (Summer 2016) – even if in name only. Her father Will Johnson was highly organised and tidied up many matters in the final decade of his life. AJK was sold to the Marr trawler-owning family around 1953. Fortunately, they have kept the name alive in the form of the modern-day AJK Cold Store business. These three AJK (combined) pictures are all in the Hessle Road area. Copyright Alec Gill.

ASIDE: Richard Nielsen (a distant relative of Amy Johnson) has a vague childhood memory of being told that the seagull on the AJK logo was called / nicknamed 'Jason'. I suspect that this notion came from someone in the family with a colourful and playful imagination.

(1) FISHING ANCESTORS: Danish Blood

Amy's ancestral roots with Hull's fishing heritage go back to her paternal grandfather and grandmother. His birth name was Anders Jorgensen, born 25 April 1852 in the Danish village of Saltofte, near Assens. He was the eldest of seven children. He emigrated to England and landed in Hull in 1869 - over three decades before Amy was born. He was a tall red-haired Dane who began as a 16-year old apprentice to a Hull smack owner (originating from Devon). Anders worked hard, used his brains, saved his money, did well for himself, obtained his master's (skipper's) ticket, bought his own fishing smack and then another.

Whilst he was busily achieving all this, he fell in love with a beautiful young English girl. On 5 January 1874, he married her at Hull's large and prestigious Holy Trinity Church – so it must have been a grand wedding[7]. Anders' sixteen-year old bride, Mary Ann Holmes, also had firm fishing roots. Her stepfather, Andrew Mudge, was a Hull trawler owner who had transferred his business from Brixham, Devon. Mary grew up in the Hessle Road area and lived in various houses within the fishing community - Somerset Street, Constable Street, Porter Street and St.George's Road. When the couple celebrated their Golden Wedding Anniversary in 1924, he told a Hull Daily Mail reporter about how he came to Hull and "...I had the good sense to choose an English girl for wife. While I am well satisfied with the success I have had in business, the best day's work I ever did was in the choice of a wife fifty years ago"[8].

On 29 January 1878, the Dane anglicised his name to Andrew Johnson[9]. The Johnsons had fourteen children (but four died young). One of their houses was No.111 Coltman Street at the Hessle Road end of this illustrious street[10]. Three of their sons - William, Eric and Herbert - became involved in the fish trade.

Image 4: Initially, I assumed that the tall, portly, top-hatted man with the neat beard was Andrew Johnson himself. Then Richard Nielsen pointed out that this central character had a wooden leg! Whoops!! But why was this photo in the Johnson archive? Richard and I now guess that Andrew Johnson might be the bearded man (to the left) with a trilby. If anyone knows what is happening in this picture, then please get in touch. For some reason, fish salesmen, merchants and dock workers are gathered around the herring boxes – some with their feet on the fish. This picture might be around 1908. Courtesy RAF Hendon Museum.

In 1881, Andrew teamed up with a fellow Scandinavian – a Norwegian called Tom Knudtzon. The two men were the first to import good-quality herring from Norway. The company prospered and they became one of the port's top fish merchants – especially in the herring and kipper trade. They registered their business as Andrew Johnson, Knudtzon & Company (Image 8). Although Tom died a couple of years later, Andrew maintained the Knudtzon name in what seems like an act of loyalty to his late friend.

Two of Andrew's siblings, Erik and Hannah, subsequently came from Denmark to join him. They both became heavily involved in the port's thriving fishing industry. Erik later had two sons - Albert and Fred - who became fish merchants. Hannah married Niels K. Nielsen – and their grandson was Ken Nielsen, Skipper of the Cape Trafalgar (H.59), who was the runner-up for the Silver Cod Trophy in 1961. Therefore, the Johnson Family were significant movers in shaping Hull's early trawling industry. Indeed, they were a powerful family in the fish trade of Victorian England and well into the 20th century.

Andrew also owned a few sail trawlers – about three smacks. Whilst researching material for my book Lost Trawlers of Hull, I came across his vessel Flower of the Forest (H.930). In May 1892, she caught fire and sank in the North Sea off Spurn Point[11]. In addition, Andrew Johnson was an inventor who patented ideas for modifications of the trawl doors used to keep the mouth of the net open underwater. He was also a man of property – owning good houses in Hull's Boulevard that he rented out to family members, business colleagues, and others.

In the 1880s, many Hull smack owners made the shift from sail to steam. Some were cautious and held off from the large investment. One prevailing Yorkshire view was that "all the profits go up in smoke – through the ship's funnel". They were wrong, and iron trawlers brought vast profits to the port. A second cousin of Amy Johnson's - Janet Scott - suggested that Andrew Johnson did not make the transition from sail to steam trawlers[12].

Some recent evidence, however, shows that he operated both smacks and steam trawlers during the 1890s[13]. Andrew was a shrewd entrepreneur who made commercial decisions that were highly profitable for his family firm and Hull's trawling industry. He invested his capital in both the sea and shore-based elements of the fish trade: catching, handling, processing and selling.

During the 1880s/1890s, AJK established the area known as Cod Farm at the very western end of St.Andrew's Fish Dock (Image 5). This vast estate covered acres of land and employed scores of people. I do not have any financial figures, but would estimate that profits from Cod Farm were the main powerhouse of the AJK empire. In essence, the firm traded in fish – buying at a low price and selling at a profit.

Briefly, white fish (cod and coley) was salted, and then dried in the wind and sun on long waist-high wire mesh tables. At the end of the long process, the salted fish was rock hard and ready for export around the world – rarely was it sold in the UK.

Image 5: This is the massive AJK Cod Farm estate at the far western end of St.Andrew's Dock (on the banks of the Humber). Note the large shed (left) with the company name emblazoned on its side. Cheap quality fish was processed at Cod Farm. Equally, when there was a glut of fish landed at Hull, the unsold, good quality fish also ended up at Cod Farm at the lowest price. There are around 70 long wire mesh tables in this sketch. This image suggests that the scale of production was massive. Courtesy 1915 NER Advertisement.

(2) FISH EMPIRE: Smelly Dock

The large AJK fish empire, between the 1880s and 1940s, divided into several distinct moneymaking branches:

(A) Trawl Fish – this division bought wet fish landed by Hull trawlers and sold it to fishmongers, hotels, restaurants, and wholesalers for distribution to fish & chip shops around the UK – like any other Hull fish merchant.

(B) Herring Merchants - this was the firm's traditional branch that Norwegian Thomas Knudtzon brought into the business back in 1881. AJK were major fish importers of prime Norwegian herring and salmon that they sold throughout Britain.

(C) Boneless Kippers - herring cured in smokehouses. On 30 August 1928, the Hull Daily Mail ran the headline: Local Firm's Initiative. The article heralded a marketing breakthrough: "a Hull firm has placed on the market a boneless kipper...The boning is done by hand by women...a wonderful development in the herring industry." An AJK spokesman is quoted as saying, "we guarantee that only really good, clean quality herring are placed on the market...I am confident it will increase the sale of kippers...we are getting repeat orders...the trade generally should experience a wonderful impetus" (p5). Coincidentally, the firm's new, profitable fish product came onto the UK market around the same time Amy was beginning her flying career in London. Is this a previously unseen reason why AJK could afford to become her sole business sponsor when no other company would fund her solo venture to Australia during The Hungry Thirties?

(D) Importers and Exporters of fish - as well as UK branches in Lowestoft, Yarmouth, Lerwick, Ayr and Anstruther, AJK had offices or agents overseas in Piraeus (Greece); Milan (Italy); the Irish Free State; Spain; Cyprus; South Africa; the USA; and Brazil (according to an office letterhead in 1934).

(E) Shipbrokers and Commission Agents – AJK probably handled vessels from Norway and elsewhere when they imported a full cargo of fish.

(F) Anglo Norwegian Fishing Company Limited – During the 1890s, Andrew Johnson took the brave step to invest his capital into steam trawlers. The company switched over from sail to steam – part of Hull's Industrial Revolution. He owned and managed at least five vessels. Each bore a Classical Greek name: Melpomene (H.1474 = 1891 sold 1896); Ceres (H.219 = 1892 sold 1905); Achilles (H.109 = 1892 sold 1906); Hermes (H.209 = 1893 sold 1899); and Socrates (H.885 = 1906 sold 1912 – Image 6). This comparatively small Hull trawling fleet only existed between 1891 and 1912. Andrew's brother-in-law, Niel K. Nielsen, often skippered these trawlers when new.

Image 6: Andrew Johnson was a leading figure in the Anglo Norwegian Steam Fishing Company. At a guess, he had a love of the ancient Classics if the naming of his fleet is anything to go by. They were named after a Greek muse, goddess, warrior, god, and philosopher. Socrates (H.885) was built specially for his firm in 1906 at Glasgow. Sold to Fleetwood (FD.163) in 1912, it was wrecked in thick fog on 7 January 1913. Courtesy Charles Ayre.

(G) Cod Farm – AJK ran a vast estate at the western limit of the St.Andrew's Fish Dock where cod and coley were salted and dried in the open air (Image 7) before being bundled together for export abroad. It was a global enterprise. In a later section, I will especially highlight the work at Cod Farm and show how its infamous reputation, within the port, affected Amy's 'shy self-conscious' teenage outlook (Chapter 7 - DREAD: Cod Farm Lasses).

Image 7: Cod Farm covered acres of land on Hull's St.Andrew's Fish Dock. AJK employed scores of people - mainly on a casual basis. Working conditions were rough and ready. Labour costs were kept very low and drew upon a pool of unskilled workers, especially during the Great Depression after the Wall Street Crash in 1929. Courtesy RAF Hendon Museum.

In effect, Danish immigrant Andrew Johnson did very well for himself in the Hull fish trade. He was not only a successful trawler owner, fish merchant and inventor, but also became a highly respected member of the Hull establishment. He was a founding member of the Fish Merchants' Protection Association; a trustee of the Widows' & Orphans' Provident Fund (for the benefit of fishing families who lost trawlermen at sea); a prominent member of the powerful Great Thornton Street Wesleyan Methodist Chapel; a leading figure in Hull's Danish Church; and a Freemason in the long-established Humber Lodge.

Andrew Johnson's eldest son was John William - known in the family as 'Will' (born 19 May 1876). He entered the fish trade with enthusiasm when he was 15 years old. Seven years later (1898), adventure-seeking Will set off to find gold in the Klondyke region of British Columbia in Canada – along with three of his mates. Despite high hopes, it was a failed expedition. All he brought back to Hull were four tiny gold nuggets. Fortunately, he had a rich father to finance his fantasies. Will nearly lost his life, but never his zeal and wanderlust[14]. Amy certainly picked up on her dad's restless spirit to travel the world in search of adventure. Like her dad, Amy would also receive direct sponsorship from the family's fish business.

Image 8: The Andrew Johnson, Knudtzon's (AJK) advert shows Amy's Danish grandfather Andrew (top). Her father Will (below) took control of the AJK empire in 1914. He alone ran the business until he sold it almost four decades later. In essence, Amy was born into a relatively wealthy, well-established fishing family that helped lay the foundations of Hull's Hessle Road Community. Courtesy 1915 NER Advertisement.

(3) BORN: A Tomboy in an Ugly Street

When Will returned home to Hull, he was ready to settle down (aged 26 years). He married Amy 'Ciss' Hodge (18) in September 1902. The Hodge Family also lived in the Hessle Road neighbourhood at No.55 St.George's Road, and in Coltman Street at No.142 and No.193 – a large bay-windowed house. Ciss was the granddaughter of William Hodge who had been Mayor of Hull in 1860 and lived in the grand Newington Hall on Anlaby Road. Although William had been a highly successful businessman and made vast profits from the port's thriving seed-crushing industry, he was too generous and gave to charities more money than he earned. To the family's disgrace, their opulent Newington Hall had to be sold. William's eldest son (Ciss's father) was a failure and flogged off more family assets. As the Hodge Family plunged down the social ranks, Amy's mother Ciss became "an insecure and twitchy young woman"[15].

Thus, in some respects, Ciss married below her family's earlier social status. She might well have spotted Will Johnson's potential prospects and motivated him to climb the social ladder. She never let go of her social aspirations for the Johnson Family to better themselves. Equally, she might have simply married for love. The newly-weds happily moved into rented property not too far from the fish docks.

Their house at No.154 St.George's Road was in the busy Hessle Road community. It was there that Amy was born on 1st July 1903 – the first of four daughters to the Johnsons. The year 1903 was, of course, when the Wright Brothers flew the first-ever human controlled, powered, heavier-than-air machine. She came into the world at the Dawn of Flight.

It is more by good fortune than design that Amy's birthplace is still standing. It survived the German bombing in two world wars. In addition, it escaped the savage bulldozers of the 'slum clearance' craze during the 1970s/80s. The home where Amy grew up received a blue plaque (Image 9) thanks to a nearby school.

Although Amy remembered No.154 as her "comfortable Yorkshire home", she disliked St.George's Road itself and described it as an "ugly suburban street". Even as a three-year old, she saw it as so ugly that she tried to flee from the place – until a neighbour carried her home "screaming and kicking and vowing vengeance"[16]. Running away was an instinctive Amy trait that showed itself almost as soon as she could walk.

Obviously, Amy was not too fond of the locality where she grew up – ashamed is not too much of an exaggeration. She was perhaps keen to escape the smell of the nearby Cod Farm. Biographer Midge Gillies (2003) stated "although St.George's Road was one of the more affluent streets, the stench of fish still hung in the air and the fishing community's traditions ruled every aspect of life" (p.12). Trawling was the most hazardous industry of its day and superstitious rituals governed many daily activities. Amy absorbed a fair degree of this local fishing folklore.

St.George's Road was also a place where Amy recalled having a rough-and-tumble when young: "My clearest memories of my father in my earliest childhood days are of him romping on the floor as a bear [no doubt from the Rocky Mountains], encouraging me to throw away my dolls and play with trains instead"[17]. She added, "to this day, there is nothing which gives me greater joy than a ride on a scenic railway or a steam yacht at the fair"[18]. No doubt, her dad also took his eldest child on the exciting and daring rides at Hull Fair every October – held only a short walk away from any of Amy's childhood homes.

Attending Hessle Road picture palaces was another father-daughter activity. Most Friday evenings they enjoyed silent movies such as adventure, romantic, historical, and grand spectacular films. "The cinema was a big influence upon Amy's life when young"[19]. When older, Amy recalled a time at a local cinema by herself when suddenly "an aeroplane appeared on the screen in a news item and, wildly excited, I sat through the whole programme twice just to see it again. For some strange reason that aeroplane appealed to me enormously. It seemed to offer the chance of escape for which I was always looking"[20]. When she moved to London, and her love life took a down turn, "...cinema-going once again became an addiction"[21]. As his eldest child, Will treated her more like a son than a daughter.

Various authors have described Amy as a 'tomboy' when growing up in the Hessle Road Fishing Community[22] and Amy herself delighted in writing about "my tomboy spirit"[23]. When Amy joined the Boulevard School, she heartily declared, "First and foremost, the school being co-educational, I rapidly developed a great interest in the opposite sex. I took a fierce pride in my popularity with my boy class-mates and became as big a tomboy as any of them"[24].

In a sense, Amy struggled with her sexual identity – perhaps thanks to her father. Boys games were far more daring, risk-taking and adventurous. She enjoyed playing cricket, for example. Girls then were more restricted and limited in what they could do. She cringed when her mother insisted upon combing and caring for her hair every day.

Amy even made herself look like a boy: "an early act of rebellion had been to cut off my hair and fasten back what was left with a hair slide. What a triumph it was! No more plaits or ribbons were possible. My mother was broken-hearted, for she had spent hours and hours brushing my hair and curling it over her fingers and she had just cause to be proud of the result. Thoughtless and selfish, I caused her still more grief and worry. I am afraid I was never a model daughter"[25].

She was the battlefield in a domestic war between her parents. Father, Will, treated her like a son and encouraged her tomboy traits; mother, Ciss, wanted her to be ladylike and always enjoyed seeing her in pretty dresses with neat curly hair. Her father won the battle in many areas.

Amy subconsciously followed the tomboy path set by Will Johnson. He had had his wild adventurous fling trekking the mountains of British Columbia in search of gold. Though he returned home disillusioned and empty handed, he unwittingly projected his restless spirit onto Amy.

Deep down, he was perhaps happy to see his first-born rocking the boat and creating havoc at school. Amy was the son he never had. She embodied her dad's ambitions. Unlike him, she did conquer the world. Amy escaped the humdrum reality that her father had to face every day in his fish dock office. When Amy achieved global fame, she did so on behalf of her father and his unfulfilled dreams.

Yet mother Ciss did have some success within Amy's mind. She projected on to her daughter her perception of the neighbourhood being 'ugly'. A place perhaps the nervous mother herself wanted to flee.

Image 9: This commemorative plaque is above the doorway of two-up, two-down No.154 St.George's Road. It declares, "Pioneer Aviator Amy Johnson (1903-1941) was born here 1st July 1903". Credit, however, must go to local pupils and teachers of the nearby Newington Primary School. In 1988 they urged Hull City Council to recognise that her birthplace was worthy of distinction – 58 years after her famous flight. Copyright Alec Gill.

(4) BRED: A Methodist Family

Edwardian Britain saw a high level of worshippers in the pews on Sundays - compared to the period after the First World War. It was part of the social norms of the time to be a good Christian. Even so, most of the Anglican congregations were from the aristocratic and upper middle-class families. Chapelgoers were largely middle-class (professionals, shopkeepers, merchants, office workers and upper working-class groups such as railway workers). Generally, the average working-class family attended a place of worship mainly for ceremonies linked to 'hatched, matched and dispatched' – birth, marriage and death.

In addition to being born into a Hessle Road fishing family, Amy had a Methodist upbringing. The Johnsons were regular attenders at their nearby Wesleyan Methodist chapel on the corner of St.George's Road and the bustling Hessle Road (Image 10). Indeed, Will and Ciss were married there on 11 September 1902[26]. The Johnson Family had a strong affinity with the place. A Methodist Conference Report described this chapel and declared, "Amy Johnson was pianist and Sunday School teacher here"[27]; but this serious mix-up may have been with Ciss. Mother and daughter were both officially called Amy Johnson. In the wake of Amy's global flying fame, someone retrospectively assumed she had taught at the Sunday School. It would have been a real feather in the cap of the chapel, if true. It was not.

Image 10: Amy's parents, Will and Ciss, were married in 1902 at this grand St.George's Wesleyan Methodist Chapel on the corner of Hessle Road. Their daughters attended the Sunday School and other activities there. Ciss loved to play the piano during services and Will was a chapel trustee long after the Johnson Family moved away from the fishing community. Courtesy Kitty Ingram.

Mother Ciss was the musician in the family. Amy recalled when young: "she played like an angel and I would lie on the sofa with my eyes shut living a wonderful secret life of my own, full of exciting adventures in which I was always the heroine and the end was always happy and satisfying. I adored fairy-tales...stuffing my head with stories of knights in shining armour, beautiful princesses with red hair who fell in love with handsome young peasants they found trespassing in the Palace woods..."[28]. Although she was a tough tomboy on the outside, inside was an innocent damsel wandering in a dreamy world.

Ciss was a very tense woman, "prone to violent mood swings" and "obsessive about the cleanliness of her house. Her one release was playing the piano...at their local Methodist chapel"[29]. Amy attended the Sunday School and was a 'scholar' of the chapel. Even after three house moves, well away from the 'smelly and ugly' St.George's Road, father Will (and probably Ciss) were still very active members of this Hessle Road establishment.

The chapel's Trustees Minutes[30] for the 4 March 1919 recorded that "J. W. Johnson acted as secretary for the meeting". On 29 January 1925, Will became permanent Secretary to the Trustees (replacing Mr. G. Tolston). During a Special Meeting of the Trustees on 5 June 1925, a big shake-up took place. The Board increased to 27 members and included "John William Johnson, Hull Fish Merchant".

As Secretary, Will would have been busily involved with this energetic place. Although the congregation was steadily diminishing from its Victorian heydays, it was a hive of activity. The chapel ran a 23-man Silver Band; Cadets of Temperance (Band of Hope) Society for youngsters who 'signed the pledge' never to drink alcohol in their lives; Clubs for Older People; Annual Charity Dinners for local residents; and the popular St.George's Brotherhood, later called the West Hull Men's Fellowship, met every Sunday afternoon with famous speakers from across the British Empire.

In 1909, the Johnson Family moved from St.George's Road to No.241 Boulevard (Image 11). Amy's paternal grandparents, Andrew and Mary Johnson in nearby Coltman Street, owned this house. The Boulevard was a tree-lined highway – more aspirational than their former 'ugly street', slightly further away from the fish dock place of work and nearer to relatives who also worked in the fish trade. Indeed, many of six-year old Amy's extended family lived in the Coltman Street and South Newington area. Thus, her kinship network mainly dwelled within Hull's fishing community. The topic of conservation around the Johnson Family table would have been about life on St.Andrew's Fish Dock, the latest trawler tragedy at sea, fishing yarns, and making fun of the Hessle Roader's folklore superstitions. Biographer David Luff (2002) mentioned that Amy and her sister Irene enjoyed their halfpenny rides on their favourite vehicle along Hessle Road – the D-tram that rattled its way from the city centre to Dairycoates[31] (Image 12).

Amy was not as keen a Methodist as were her parents. She even began to have doubts as a schoolgirl. Biographer Constance Babington Smith (1967) conducted a wide range of interviews with Amy's contemporaries from her childhood to her death. She noted that "even during Amy's schooldays, although she attended Methodist meetings, was never an enthusiastic chapel-goer. She never showed any religious zeal" (p30).

In April 1921, when Amy was 17 years old, she wrote a piece for The Boulevardian school publication called New Year's Day in China[32]. This interesting two-page article highlighted "the family gods" and how "many loud and brilliant fireworks are sent up by the priest, and long strings of fire-crackers are lighted on the threshold of the houses, it being believed that all evil spirits will thus be kept away during the ensuing year" (p27). As well as paying off arrears to people, "debts to the gods are also settled by the more devout Chinese by means of gifts and prayers" (p28). This writing seems to reflect a youthful interest in New Year superstitions.

By October 1922, the gulf in Amy's beliefs widened even more when she left Hull and began at Sheffield University: "along with her friend Winifred [Irving], Amy also lost her Methodist childhood beliefs"[33]. I believe that, in the absence of her Methodism, folklore beliefs - picked from the local Hessle Road fishing families and school friends - filled her spiritual vacuum. The more risks Amy took in her flying life, the more she drew upon a primitive belief system of good omens and protection by the gods. Gillies (2003) stated, "Although she did not class herself as religious...she spoke often of the guardian angels who were watching over her throughout her journey to Australia" (p148). Then, when the Second World War came along, Amy coped with the bombing in "a fatalistic way"[34] – much like the Hull trawlermen during a winter voyage to the Arctic waters.

Image 11: No.241 Boulevard was the Johnson Family's next home (1909) – opposite the Western Library. Visiting relatives recalled that the household was always "lively and full of fun". This jollity was probably due more to Will's exuberance than Ciss' mood swings. Copyright Alec Gill.

Around 1910/11, the Johnsons were moving house yet again. This third move was to No.48 Alliance Avenue (Image 13). Their new home was a few streets away in a north-westerly direction, over the Anlaby Road level-crossing and beyond West Park, in the North Newington district. Each flit reflected an improvement in the family's financial circumstances. Each new home was slightly bigger and the neighbourhoods a little more modern than before. At the same time, the Johnsons were geographically distancing themselves further away from Cod Farm. Young-at-heart Will enjoyed roaring to his fish dock office on his latest motor-bike. Soon after the move to Alliance Avenue, Ciss gave birth to Molly in 1912 – the third daughter in the family.

As the storm clouds gathered before the start of the Great War, Will Johnson took over the running of AJK's fish business. At thirty-eight, he was too old to serve in the conflict and his role in the vital food industry was a 'reserve occupation' – essential for national survival. He was on a committee of the HFVOA (Hull Fishing Vessel Owners' Association) who liaised with the Ministry of Food and this involved Will on frequent and enjoyable trips to London. After WWI, the Johnsons went on to become a two-car household[35]. Although retired Grandfather Andrew spent lots of time in Bridlington, he still kept a watchful eye on the Cod Farm business and its profit levels for the next twenty years.

No matter where on earth any of us are born and bred, that initial place leaves an indelible impression for the rest of one's life. Regardless of what anyone goes on to achieve, we can never deny our earliest beginnings. Amy grew up in the streets of the Hessle Road Fishing Community and mixed with other children in the neighbourhood and school. She had cousins whose families were directly engaged in Hull's fish trade. She was a Hessle Road girl in every sense – through and through – and a tomboy at that!

Image 12: This is a great Hessle Road picture (c.1905) - it combined two elements of Amy's life in one. This is the D-tram from Hull city centre to Dairycoates. Amy and her sister Irene enjoyed riding along Hessle Road for half-a-penny. In addition, the large building to the left is the St.George's Wesleyan Methodist chapel where father Will was an active, long-term Trustee and Committee Secretary. Courtesy Chris Ketchell.

(5) FORMATIVE YEARS: Boulevard Scholar

Amy grew up in the fishing community, but did not attend any of the nearby Infant or Junior Board Schools with local pupils. Coming from a relatively wealthy background, her mother was eager to send her to private schools in the Newington district (between 1908 and 1915) - "none of which were of a high academic standard"[36]. Instead, girls were expected to become young ladies – a stepping-stone to a good marriage and child rearing.

Janet Robson (née McDougall) was a fellow pupil of Amy's at one of these private schools between 1907 and 1912. Eversleigh House was at No.557 Anlaby Road near the corner of Glencoe Street. Two sisters ran the school on a strict basis. Amy was "a very high spirited little girl and I remember her dancing on the teacher's desk when she was out of the room"[37].

The Headmistress, Ada Knowles, was extremely keen on geography and often pointed to a world map on the wall. The girl who could correctly name the town, river, country or capital city in question received a good mark. Janet and Amy competed to give the right answer. Upon reflection, Janet wondered if Miss Knowles' geography lessons planted the seeds of interest in faraway places – resulting in Amy's flights to Australia and elsewhere around the globe.

Certainly, the geography lessons were important, but I would quickly add that Amy already had a deep, pre-school interest in faraway places. After all, her grandfather was from Denmark, her own father often talked about his youthful adventures in Canada, and her family's wealth arose from trading with exotic places such as Brazil, Cyprus, USA, Norway, South Africa and Greece. Thus, during mealtimes at the Johnson household, the air would be full of such colourful place names.

Tomboy Amy certainly proved to be 'a square peg in a round hole' at these private establishments. Her time there was not a success from their point of view. Amy herself had a more positive perspective. She admitted to being "headstrong and probably spoilt, with an insatiable desire to know everything...I brought homework home with me for the sheer love of doing it, and gave my teacher hours and hours of extra work correcting interminable sums of algebra...when I finally left private school...I could boast of a smattering of physiology, algebra, geometry, trigonometry and biology"[38].

Image 13: Amy Johnson's third family home at No.48 Alliance Avenue. Luff (2002) pointed out "Amy and her sisters lived in a warm, close-knit family atmosphere with a constant stream of aunts, uncles, grandparents and friends visiting their home in Alliance Avenue" (p36). Copyright Alec Gill.

Amy's mind sparked off in many directions. It was like a sponge, eager to soak up new information. She was a highly gifted child, too quick for her plodding teachers, set in their ways of teaching the same topic year after year. She danced on desks because her dull tutors and their rigid rules bored her. She was too lively for their limitations: "a bit of a handful" was the typical Hull expression to describe such a child - not the well-mannered material that Eversleigh liked to produce.

Amy's parents finally recognised that sending her to more private schools was a waste of time and money. Perhaps out of desperation, exasperation (or even as a punishment), they decided to send her to the nearby Boulevard Municipal Secondary School for the final years of her education (1915-1922 – Image 14). They thought she might even benefit from mixing with ordinary children of her own age. Her parents were running out of viable options for headstrong Amy. Fate, however, did not work in their favour.

Image 14: Amy's splendid school stood at the northern end of the Boulevard near Anlaby Road and West Park. Her first impression of the interior was that it was a "gloomy cheerless place". As a scholar there, she possibly exhibited her fiery nature to 'cheer up' the dull place. When Amy started at the Boulevard, Will presented her with a special Waterman's Ideal fountain pen. She kept this with her over the years and used this school memento during her flight to Australia. Courtesy Boulevard Booklet.

Due to an administrative mix-up, Amy was older than her classmates and more academically advanced. History was about to repeat itself, she was bored and recalled, "I hardly had to do any work to be easily at the top of the class. I therefore developed excessively lazy habits"[39].

Amy had masses of mental and physical energy to burn. Within the school context, that was enough to label her as a "rebel" and troublemaker. She was very keen on a wide range of sports: swimming, athletics, trapeze, springboard, hockey (Image 15), and especially cricket. She was the only girl in the school who could bowl over arm and do so with great skill. In addition to sport at school, Amy considered herself lucky to go sometimes "three times a week to the gymnasium of the Young People's Institute" in George Street within Hull's City Centre[40].

Image 15: Although Amy was at the Boulevard School for several years, she never became a prefect. This was probably due to her lateness and rebellious tomboy nature. In April 1921, Amy was Secretary of the Girls' Hockey team (back row, far right) and a member of the Girls' Cricket Committee. Courtesy RAF Hendon Museum.

Elizabeth Grey (1966) explored, more than most biographers, Amy's Boulevard schooldays by interviewing former classmates. Grey described how "the headmistress was her deadly enemy" (p5). This was because of Amy's many pranks and her frequent lateness. She easily became bored with certain topics, was late for these classes, and often sent to the headmistress for punishment. Indeed, Amy became an inventive fibber in providing reasons for her unpunctual behaviour. She admitted, "I became a most fertile liar…and how many excuses I have invented! The passing of the years has not in any way dulled this ability"[41]. This deceptive trait proved useful throughout her life when laying a false trail to reporters and friends about her real place of birth and family origins. She found it easy, and perhaps fun, to make up fanciful stories about her background.

As in many areas of her life, she went against the flow and was an anarchist, working hard at unladylike subjects such as trigonometry[42]. Nevertheless, she saw her time at the Boulevard as having "occupied seven of the most valuable years of my life – I was at school until I was nineteen – I will relate...incidents which seem to have had the most influence on my future career"[43]. She then focused upon two specific events – to which most biographers gave full attention. These were (A) her straw-hat rebellion and (B) the cricket ball in her mouth accident. I also concentrate on these landmark events, but with a different twist.

(A) The Revolt of the Straw-hat Brigade: There is no exact date, but this incident revolved around the wearing a straw boater as part of the girls' school uniform. "I loathed and detested the ugly straw 'bangers' we had to wear"[44]. Amy hated it so much she devised a strategy to confront authority in the hope of getting it changed. She preferred the Panama hat worn by girls in another school. If she did not like something, rather than passively accepting, she took it upon herself to alter the situation.

Her first step, as class 'ringleader', was to convince her mates to each turn up at school sporting a Panama hat, which they all agreed to do on a certain day. It was a courageous plan and showed that Amy had the imagination to formulate a plot and the guts to put it into action. On the other hand, it could have been her ego working overtime. Was her idea a convoluted form of attention-seeking behaviour? In the event, Amy was the only girl to arrive wearing the 'illegal hat'. Her friends let her down. The full weight of school discipline fell upon her head alone.

The fact that her 'friends' did not stick by her and support her headgear protest, made Amy feel not only a fool, but betrayed. It seemed that everyone was laughing at her behind her back. She had stuck to the originally agreed plan (her plan, that is), but others had conspired against her and decided to conform to the school rules instead. Institutions rarely respect individuality. She must be 'taught a lesson'.

Amy described how "I was ordered to stand outside the headmistress's door in my Panama hat until she should come out and see me. The corridor...being a very busy one, almost everyone in the school had the chance to pass me and twit me for being in such an undignified position"[45]. So in effect, it was a blow to her pride and self-image. Tomboy Johnson easily took the physical punishment by the teachers, but being laughed at and ridiculed by friends left a deep and lasting wound upon the inner Amy.

(B) Cricket Ball: The second and more serious blow to her pride and extrovert antics occurred on the cricket pitch during the summer of 1917. The powerful impact this had upon Amy is best described in her own dramatic words. This is a long (but telling) extract, in which she openly declared,

> "A tiny incident which seems to have had a much more far-reaching effect on my life was a cricket ball hitting me full in the face and breaking one of my front teeth. Dentistry in those days not being what it is now, I went through my school days with this broken tooth, and from thence onward spoke with a slight lisp. There was no doubt that my looks were seriously impaired from the point of view of my male class-mates and, from being the ringleader, I quickly adopted a most unfortunate inferiority complex which followed me through the rest of my life and probably had a most important bearing on my future actions and character. The boys made fun of me, and I was so self-conscious that I avoided them. I became introspective and withdrew farther and farther into a protective shell of my own making. Instead of spending so much time on games, I acquired the liking for long cycle rides. Many and many a time I played truant from school, cycling for miles into the country with some sandwiches or biscuits in my pocket, which I had stolen from the larder. I also tried again and again to run away. With what object, I do not know. Probably just that craving for "Escape" which has driven many a young man and woman in the last few years to run away from "they don't quite know what" to find "something of which they have no clear idea". My poor parents must have been sorely tried by my inexplicable conduct, but there it was. Feeling myself an outcast, I sought always the hermit's life, and kept away from school as much as possible."[46]

On two significant occasions, the children of Hessle Road mocked her. Former friends now 'took the micky' out of her. It was a major turning point in Amy's life. She would never be the same again. Her negative Boulevard experience shaped her personality forever.

Dr. ALEC GILL MBE
Author, Historian, Film-maker
T: 01482.225009
M: 07786_582195
E: alec.gill@hotmail.co.uk

THREE different VERSIONS of my AMY JOHNSON Book are now available between £2.99 and £23.00
BEST ordered ONLINE – rapid service.

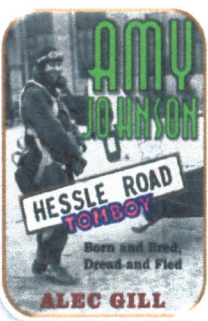

eBook / Kindle Version available via AMAZON for £2.99:
https://www.amazon.co.uk/AMY-JOHNSON-Community-Trawling-Heritage-ebook/dp/B015ST0GEM

B&W: Black & White Print Version for £8.00
https://www.amazon.co.uk/AMY-JOHNSON-Community-Trawling-Heritage/dp/1535109947
ISBN: 978-1535109949

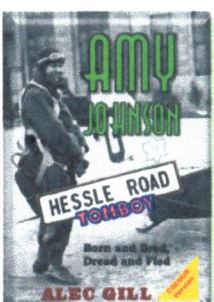

Colour Print Version for £23.00
https://www.amazon.co.uk/AMY-JOHNSON-Community-Trawling-Heritage/dp/153306203X/
ISBN-13: 978-1533062031

DR. ALEC GILL MBE
AUTHOR, HISTORIAN, FILM-MAKER
T: 01482. 225009
M: 07788. 552190
E: alec_gill@hotmail.co.uk

THREE different VERSIONS of my AMY JOHNSON Book
are now available between £2.99 and £23.00
BEST ordered ONLINE – rapid service.

eBook / Kindle Version available via
AMAZON for £2.99:
https://www.amazon.co.uk/AMY-
JOHNSON-Community-Travelling-
Heritage-ebook/dp/B01SSTOCEM

B&W: Black & White
Print Version for £8.00
https://www.amazon.co.uk/AMY-
JOHNSON-Community-Travelling-
Heritage/dp/1545109947
ISBN: 978-1545109649

Colour Print Version for £23.00
https://www.amazon.co.uk/AMY-
JOHNSON-Community-Travelling-
Heritage/dp/1533052794/
ISBN-13: 978-1533052931

(6) FLED: Have Bike, Will Escape

Out of favour with her peers, Amy was now alone and vulnerable. After being such a popular pupil, she withdrew and took to a solo pursuit. Escape was essential for the sake of survival. She chose to be semi-isolated via a mechanical means: the unladylike contraption of the then modern-day bicycle. Thus, in that awful summer of 1917, desperate to escape the torments, she took to her bike and rode alone around the villages in the foothills of the Yorkshire Wolds. "Another pleasant ride, to the west of Hull, began past the turning to the fish dock. Soon the Hessle Road led out on to the flats alongside the Humber"[47] – she liked to cycle out to Brough, about 10 miles away.

The First World War was still raging when Amy "discovered the Blackburn Aircraft Factory at nearby Brough...and began to cycle out by herself to watch from a distance the planes being assembled"[48]. This was a brand new factory, established for the war effort in 1916, to design, build and test seaplanes for the British forces.

Hull-based biographer Bob Finch (1989) saw Amy's cycling and love of lonely outdoor pursuits as having a very high priority and that her desire for the open-air life began at the Boulevard. He outlined many examples from her adult activities such as camping, walking, gliding, horse riding, skiing and sailing - "[these] had a long lineage stretching back to the times when she was a teenager and went on bicycling tours". He concluded, "For Amy, flying was just another outdoor activity"[49] – especially in an open cockpit.

Fourteen years of age is a critical and sensitive time of change for anyone. A young person's body image has a profound impact upon shaping the adult personality. The aftermath of the headgear and cricketing incidents left Amy with two interlocked personality traits. These were not only firmly entrenched during her formative years at the Boulevard School, but also set a lifelong pattern for how she would deal with deep emotional problems in the future.

The first trait was that Amy loathed it when people laughed at her behind her back[50]. These embarrassing social situations then triggered a second trait: the urgent need to escape, usually, via some machine. I am thinking specifically of the bicycle and aeroplane. Both these mechanisms were, at the time, considered unladylike. Indeed, that too became an added (third) motivation to persist in an eccentric course of action. Perhaps the 'unladylike' element was simply a reinforcement of her earlier tomboy trait – encouraged by her (son-less) father. In essence, the dread of social ridicule sparked off the impulse to flee via a mechanical means of escape.

Image 17: The Endike Community Centre, 21st Avenue in North Hull has a splendid wall mural with a variety of colourful local scenes from history. Amongst the many pictures by Hull artist Gordon Gledhill, is one of Amy and her famous Gipsy Moth plane Jason. Interestingly, Amy's first-ever flight was from a farmer's field just off Endike Lane in November 1926 on a five-minute trip for five shillings with her sister Molly. She found the 'pleasure flip' a disappointment and "my hair was blown into a tangled mass" (Johnson, 1938, p141). Copyright Gordon Gledhill.

Amy's personal battles took place against the backdrop of the Great War. It seems, however, the Johnson Family were not as directly or personally affected as many families in the port. She recalled that the "war days left very little mark on my life"[51]. Regarding the Zeppelin raids on Hull, they "were grand fun as we could stay up and play games and have hot chocolate in the early hours of the morning when we ought to be in bed"[52]. She loved to dash outside to watch the raids – the nearby docks were a strategic target for the Germans.

The war, paradoxically, was a prosperous time for the AJK fish business. Norway remained neutral during World War One and so their lucrative herring trade was not interrupted too much. One sign of the Johnson Family's prosperity was that, in 1918, they moved even further away from the fish docks to No.85 Park Avenue (Image 18)[53]. This was by far the most expensive home the family had so far. It was a larger property in a more respectable neighbourhood.

Ciss was certainly pleased with her new nest. Soon after moving in, she gave birth to their fourth and final child, not the much-wanted boy, but another girl – Betty, born May 1919[54]. Will was also happy. He bought himself a brand new Standard Salon car because he had further to travel into his offices on St.Andrew's Fish Dock. It is worth pointing out that the wealthy Johnsons were not unique in moving away from the Hessle Road area when finances allowed. Many local fishing family lads, when they became big-earning skippers, liked to move to the villages west of Hull such as Anlaby, Hessle, Kirkella or Swanland.

Mr and Mrs Johnson were delighted with their new abode, but their first daughter grew more and more isolated from the world around her. One former teacher at the Boulevard stated that, by 1919, 'lone wolf' Amy had no intimate friends, nor did she belong to any cliques[55]. She only spent four years residing in The Avenues – at a most depressing and withdrawn time in her life. Her schooldays ended in July 1922.

Three months later, nineteen-year old Amy left Hull to study at Sheffield University. Thereafter, her return visits to the city were only for brief spells - her formative years were now behind her. In effect, Amy Johnson was a born and bred Hessle Road girl and her fishing family roots influenced her for the rest of her life. She had moved physically from Hessle Road, but its culture still saturated the whole of the Johnson Family. They were, after all, large-scale fish merchants and their lifestyle and financial wellbeing depended totally upon the vicissitudes of Hull's fish trade. Fish and the Johnsons were enmeshed. It was in their blood.

As a psychologist, I have already highlighted some of Amy's personality traits such as: her tomboy behaviour; leanings toward superstitious beliefs; her dread of being laughed at behind her back; her flight into hermit-like isolation; and her preference for unladylike means of escape. I now intend to expand upon a sixth trait in her nature – her schoolgirl delight in being "a fertile liar". She enjoyed telling fibs. It was amusing to enter the world of make-believe. Playful fairy-tales and fantasy were fun.

Amy found it easy to be in a state of denial about her own origins. She erected an effective (and enduring) façade around her true roots. During her undergraduate years, she deliberately lied to fellow students and tutors that she came from Bridlington; when, in fact, this was merely a Johnson family holiday home and the place where her paternal grandparents had retired[56]. In Amy's mind, it was easy to blank out Hull, Hessle Road and its Fish Dock. But why deny and lie about her place of birth? It was not just a reflection of her mother's social snobbery about working-class neighbourhoods. There were other unspoken factors at work.

The primary purpose of her smoke-and-mirrors deception I believe was to save her from a dreaded, embarrassing social situation. She did not want anyone outside her immediate family to know that her father owned and profited from the smelly and awful Cod Farm business. She did not mind being a tomboy, but she would have been horrified if labelled, by association, as a type of 'Cod Farm Lass'.

Image 18: No.85 Park Avenue was the Johnson's fourth and final family home in Hull. The Avenues Residents' Association arranged for there to be a green plaque to celebrate Amy Johnson as a one-time resident there. I would quibble, however, over the point that she did not actually live there between 1918 and 1927. From 1922 until 1925, Amy was an undergraduate at Sheffield University. Even during the summer vacation in 1925, her mother actually banned Amy from staying there although the house was empty – probably because Amy had a foreigner lover at that time and Ciss was not keen for them to 'use' the house! Copyright Alec Gill.

(7) DREAD: Cod Farm Lasses

I have spoken to various distant relatives of Amy Johnson's and not one knew anything about AJK's massive Cod Farm business – they only recalled about the herrings and sailing smacks. I suspect there was some degree of embarrassment that the Johnson Family owned Cod Farm. It had a foul reputation within the port of Hull. AJK bought cheap fish, used cheap methods to process it, and employed cheap labour (especially women) to do the messy work. Consequently, the Cod Farm lasses were the butt of fish dock humour. A polite phrase to describe them was as "a fun-loving bunch of workers"[57]. On my DvD called Hull Fish Dock[58], Kit Scrubber boss, Sid Page, told how his men "used to pull the leg of the Cod Farm lasses". British Transport Police Officer Tom Nendick remembered one lady in particular: Big Mary and all her Crowd - "You daren't say anything to them because they'd roll their sleeves up and you got one" – a thump!

The Cod Farm women's pranks were renowned, especially if a cheeky young lad started work at Cod Farm and fancied himself as 'God's gift to women'. A bunch of the lasses soon initiated him and put him in his place. More than once people said, "I daren't tell you what they got up to!" – but eventually did. Other tales involved young, naïve telegraph boys who feared being sent to the isolated Cod Farm on a bike with a message: they risked being caught by the lasses – something about a milk bottle. Their colourful language was described as "rude, crude and lewd".

Cod Farm was a rough and ready place to work. It was hard and filthy graft – not to mention the smell (but I will, later). Apart from some regular male workers, most labourers were hired on a casual basis, especially women (Image 19).

Demand for labour varied from time to time. The supply of fish depended upon a number of variable factors. It had to be cheap. It had to be coley or cod (haddock was unsuitable due to its flesh not being good for the salting process – it was better smoked). It had to be huge, a couple of feet in length, because smaller fish involved too much work. It had to be in vast quantities to justify employing a large number of casual workers – only bulk deliveries made the process worthwhile.

Image 19: Cod Farm Lasses were notorious. They were a tough breed with a very hard job to do. My belief is that middle-class Amy dreaded the possibility of being labelled or stigmatised as a 'Cod Farm Lass'. After all, the link between them was not difficult to make. Her Johnson Family owned the fish business. This may be one explanation why she distanced and dissociated herself from Hull, Hessle Road and, especially, the Fish Dock. Courtesy Ros Holt.

The AJK buyers on the market were only interested when there was a glut of fish available at a low price or going for next to nothing; then they were happy and bought a thousand kits at a time. If AJK had a massive backlog of orders to fulfil and demand for dried fish was high, they sometimes negotiated with an Icelandic trawler skipper-owner to buy his whole catch outright. The bought kits of fish were then carted around to the Cod Farm estate for processing. The fish was dumped high and piled in large sheds.

I am indebted to Albert Davis for recollecting his working days at Cod Farm during the 1950s[59]. When he started working there, aged 15, his first job was to chop the head off each fish and throw it into an empty kit. When full of heads and offal, a load of kits were sold to Hull Fish Meal[60] – there was an unwritten policy on the dock of 'waste not, want not'.

The next stage was done by the Cod Farm lasses. They were highly skilled at splitting each fish. Many were former Scottish herring girls who had married and settled in Hull – they often brought their daughters and friends into the job. After the splitting of each cod or coley, it was laid, spread-eagled on waist-high wire mesh tables. Each opened fish formed a triangular shape - roughly, like an isosceles. These were laid alternately side by side in 'blocks' along the tables – row after row after row – with little waste space in between (Image 21). The chicken wire enabled the wind to blow and dry both sides of the fish. This technique was known as 'field drying'.

Image 20: Notice the "A.J.K. & Co" hand-painted lettering on the sides of the wooden kits of fish. It looks like the men, with their oilskin aprons, have just taken delivery of a batch of fish from the market. This picture was probably taken during the Edwardian period c.1908. Courtesy RAF Hendon Museum.

Salting came next – lots of it. Salt drew out the water. After a day or so, the carcass was pressed, water squeezed out and re-salted. The next factor was time – a few weeks perhaps (depending upon the weather) – during which the fish were turned over and salted again. The work was monotonous and very hard on the skin and women's hands. The weather, of course, was vital to the whole cycle - the wind and the sun. The long-winded and formal title of Cod Farm was 'The Salt Fish Industry's Hull Sun Fish Drying Farm'.

Hull is not celebrated for its sunny climate, but there was plenty of wind blowing along the banks of the Humber. Gillies (2003) studied Will

Johnson's diaries in detail. At one point, she queried his wearisome, almost obsessive, noting down of the weather day after day. This preoccupation with the weather was because his profit margins depended on it. The less it rained, the quicker the next batch of dried fish was ready for export.

Image 21: This picture shows the blocks of fish laid end-to-end in the field drying method. These two women are laying out the fish on the wire-mesh tables. It gives some idea of the large scale of fish processing on Cod Farm. Courtesy Steve Radge

Whenever heavy prolonged rainfall looked imminent, massive tarpaulin sheets were dragged over the fish. Sticking above some tables were wood supports for the sheets. The canopy kept the tables dry and allowed the wind to continue to blow around the fish.

I have already made mention of the smell generated by Cod Farm; but the nearby Hull Fish Meal factory was also a culprit. The port's main railway line into Hull's Paragon Station took passengers close by St.Andrew's Fish Dock. Kingston-upon-Hull officials have never been happy with the city's national reputation as 'a northern fishing village'.

Another serious issue was vermin. Rats as big as cats were drawn to the piles of fish like metal to a magnet. Albert Davis told how "we chased rats with shovels until we had them cornered; then, a few slaps with the shovel had to be administered to cure that problem". Medically, rats and their urine can cause life-threatening Weil's disease (Leptospirosis)[61].

I also wondered about worms and maggots within the fish themselves. Perhaps the salt killed them off, but I am no authority. Another thought was: what about seagulls? All the piles of smelly fish must have attracted them from miles around. They are scavengers and not noted for being the cleanest of birds. Perhaps that is where the canvas sheeting was useful too? In the summer months, big blue bottles and other insects must have been yet another problem.

Albert Davis also pointed out that, after there had been a big delivery of fish, it might take several days to sort, so it piled up in the sheds "...and by, did it stink!" he added. The fish at the bottom of the pile, after being crushed by the weight above "was in no fit state for human consumption in my opinion...Yet after splitting and salting, it looked no different from the other fresher salted fish". One trawler owner confided, "in the olden days, the fish that was going off a bit, was salted, because it preserved it a bit better"[62]. It was a filthy, smelly, messy, low-grade and low-paid job. Working conditions for the AJK staff were not ideal or hygienic.

Image 22: Cod Farm women preparing another batch of fish in one of the AJK sheds. Many of these AJK pictures were kindly supplied by the RAF Museum in Hendon. Before he died in 1963, Will Johnson donated his diaries, documents, and fish dock photos to this London archive for posterity. I am not sure how this came about. Nothing seems to have been freely donated to any public body in his hometown – but more of this aspect in my final chapter (Image 43). Courtesy RAF Hendon Museum.

By the end of the full drying process, the salted fish was so hard "you could use one as a cricket bat and have a game on the dock", said Brian Hodgins - whose work took him onto Cod Farm. The board-like fish was bundled into bales, strapped together (Image 23), carted to the cargo docks (like any normal freight), and exported around the world. Shipping costs were also low.

AJK's Cod Farm profits came primarily from the export trade. British consumers never really acquired a taste for dried salted fish. Nevertheless, the Johnsons built up a vast and lucrative market in Catholic Europe. There was a preference to eat fish on Fridays and particularly Good Friday. Believers are not allowed to eat meat on the day Christ was Crucified, but fish was fine and encouraged as a good food appropriate to that holy day.

Dried fish also sold in countries with a hot climate and poor population: West and South Africa, Egypt, India, Caribbean Islands, South America and elsewhere. It was easy to transport and did not need to be frozen. Perhaps because both founders of AJK (Jorgensen and Knudtzon) were foreigners themselves (Danish and Norwegian), they were more aware of overseas markets.

Foreign customers, after taking the fish home, soaked it in water for a few days to extract the salt – changing the water from time to time. The de-salted fish was then cooked in the normal way. Apparently, the taste is delicious and very enjoyable. AJK had a good business model that catered for the needs of the growing populations of the Commonwealth and around the world. This lucrative enterprise funded the Johnson Family in their personal dreams and ambitions.

AJK, however, did not have a monopoly in the cured fish business. In pre-WWII Hull, they had competition from firms such as British Fish Curing, Hampshire Birrell, and A & M Smith. When it came to exporting, prices had to be low and quality high. Nevertheless, AJK were the 'big boys' on the fish dock in that specialist field.

Given the notorious reputation pinned onto Cod Farm, it is perhaps understandable that Amy was keen to avoid the potential risk of being associated with the smelly place owned by her dad. As she, and perhaps her three sisters, got older, they became uncomfortable by this Johnson Family link and therefore distanced themselves from the Hessle Road Community. Grey (1966) described Amy the Boulevard scholar as "a shy self-conscious girl" (p5). It seems that teenage Amy was highly insecure, very self-aware and uncomfortable at divulging too much about her private, family life.

Deeper than that, her biggest nightmare may well have been the fear of association with the Cod Farm Lasses – with all the negative connotations and derision that it would involve for her. This might explain the denial of her home city, never mind her community birthplace. Cod Farm, I would suggest, was another example of Amy's dread-and-fled situations.

Can anyone get away from whom they really are and be happy? Amy might have taken some degree of pride in hoodwinking outsiders about her origins; but, as Abraham Lincoln once said, "You can fool all the people some of the time, and some of the people all the time, but you cannot fool all the people all the time".

Never mind fooling other people, what about trying to fool oneself. The facts are the facts - we each are born and raised somewhere. Having a Danish grandfather, she should have followed the advice from the Shakespearian play "Hamlet: Prince of Denmark":

"This above all, to thine own self be true
And it must follow, as the night the day
Thou canst not then be false to any man (one)"[63].

Image 23: Rock hard, salted fish did not require any refrigeration. Transport costs were kept to a minimum because it was moved around like an ordinary export consignment – baled into bundles. Courtesy RAF Hendon Museum.

(8) ROOTLESS and RESTLESS:
Alone in London

Given Amy's rootless state of mind, it is perhaps not so surprising to find that after leaving school, she entered a restless phase that lasted the best part of a decade. Her time at the Boulevard school ended when she was nineteen years old in 1922. She explained her late leaving age was because of another youthful rebellion against her parents – when she cut her own hair very short. "My father, stern and just, decreed as my punishment that I must stay at school until I had grown my hair again"[64]. Whilst there, she applied herself to passing the Oxford and Cambridge Senior examinations; but "I had not the slightest idea what to do with my life"[65]. During her limbo phase, for a brief time in 1922, she became "a teacher in the Primary Department of the Wesleyan Methodist Chapel, Princes Avenue" around the corner from Park Avenue[66].

Being at a loose end, she eventually agreed to apply for a place at Sheffield University. Before actually getting there, she had visions of studying French, English, Latin, and / or Modern History[67]. She expected then to obtain a teaching diploma – after all, she got on well with the children at the chapel. Fortunately, she had a father who was making enough profit from his AJK business to pander to his daughter's woolliness and illusions.

Her parents must have hoped she would find her feet at university, settle down to a normal life and perhaps find a 'normal' boyfriend – Amy did not do 'normal'. Will and Ciss were very unhappy about a 'strange' boyfriend she was besotted with during her Avenues phase. They could see he was bad news. Would she listen? Ironically, their very disapproval may have made him more attractive to her! Her parents' fears proved right in the long run. The problem for Amy was that 'the long run' overshadowed several years of her life. More of her boyfriend later.

Sheffield was not a rip-roaring academic success. Amy even failed her initial entrance exam. She faced humiliation at the very first hurdle and would have to return home immediately to Hull. Fortunately, a sympathetic Dean came to her rescue and offered her Economics. She took it – Hobson's choice!

Amy enjoyed the social side of student life, however, and she was 'a typical student' in many ways. She learnt the latest dances (foxtrot, tango, Charleston, and waltzes), took part in lively debates, rag week events, dramatics, and gained a couple of close lifelong friends (particularly Winifred Irving). She failed, though, to get on with most of her roommates and "In my three years at Sheffield, I changed [digs] probably thirty times"[68]. Unable to mix with fellow students, she 'escaped' once more into her hermit role and "attended the minimum of lectures" - repeating her Boulevard pattern. It was clear to her parents that a fourth year of study for a teaching diploma was a complete waste of time and money. Amy was no academic. How could she teach when she was not too keen on learning?

In the Summer of 1925, Amy returned to No.85 Park Avenue disappointed – with no glittering prizes. She only managed a mediocre degree in Economics, not exactly a topic close to her heart. Her parents had perhaps over protected their daughter much too much during her school and university life. When it came to facing the world of work, Amy was ill equipped. Reluctantly, she undertook a month-long typing course at Wood's Shorthand & Secretarial College, not too far from The Avenues, on Spring Bank near the Botanic level crossings. At the same time, she applied for office-work[69]. Tongue-in-cheek, she mentioned how "letter after letter was sent, setting out a grand total of my qualifications. It was inexplicable to me that, with my education, I was not immediately given a directorship"[70] – I love her sense of humour.

Eventually, she found employment as a secretary with an accountant in Hull's Old Town. In a sense, she was 'over-qualified' to undertake such a lowly job. There were tensions with other female office workers – they made fun of her many typing errors. Outcast Amy bluntly stated, "I hated the job. I was terribly unhappy"[71].

Interestingly, it never seems to have occurred to Will or Amy that she did some secretarial work on St.Andrew's Fish Dock – in his or a colleague's office. Will had many trawler owner contacts – it would have been easy for him to fix up a job for his wayward daughter. Women did happily do secretarial work on the fish dock. Each morning, father and daughter could have driven into work together from Park Avenue. It could have been a good temporary job whilst she found her feet, providing a degree of stability, and an income. Perhaps it was that Will, and especially Ciss, did not want any of their four girls to be involved with the fish trade. Even within management, there was perhaps snobbery toward the dirty, smelly job of feeding the nation.

Image 24: In this AJK office picture (c.1914), first on the left looks like Andrew Johnson along with his three sons: Will, Eric (stood) and Herbert. Had Amy worked there, she might have made them look a bit happier in their work! Courtesy RAF Hendon Museum.

After three years of relative freedom and fun at university, Amy found office work extremely stifling. After only three months at the accountants, she fainted and the boss sent her home. There was no psychiatric diagnosis – they kept it out of medical hands; but her 'nervous breakdown' resulted in her leaving the job. All her desperate parents could think to do was send her to Bournemouth to recuperate at the Eddison's Guest House. It had worked earlier for their second daughter Irene during one of her violent mood swings, so Amy was sent there too.

The Eddisons brought Amy breakfast in bed every morning and she moped around; but she became bored in the seaside resort during the winter of November 1925[72]. She had little to do except dwell upon her recent failures, pine over her long-term (indifferent) boyfriend, and write many letters to him. After a few dismal weeks, she was back at Park Avenue.

Family friction at home did not help Amy's mental wellbeing. She was at loggerheads with both parents regarding her expensive, wasteful three years in academia; her failure to become a teacher; her inability to earn a living or pay her way at home; and, worst of all, she was still in a shaky relationship with a man who was eight years older than her. On top of that, she was afraid to confess to her parents that she was deeply in debt to the sum of £50 - this was a result of her carefree student days (valued at £2700 in 2015). The Avenues were not the happiest of times for Amy. She was on a downward spiral, and she still had a long way to go before she hit rock bottom.

In stark contrast to Amy's calamities, her younger sister Irene was 'success personified' in every area of her sparkling life. Amy once said, "I suppose I'm hatefully jealous of her – she always outshines me"[73]. Irene, for example, had a wider circle of lively friends; was considered much more beautiful than Amy; was vivacious and charming; had a lavish 21st birthday party (Amy struggled financially to buy her a gift); had a wardrobe of new clothes; was physically fitter; drove her own car; had a well-paid job with good prospects; wrote articles for the Hull Evening News; played tennis regularly; and was successful in amateur dramatics. Will and Ciss adored her. They made allowances for her fluctuating temperament (perhaps an innate trait from the Hodge side of the family).

In March 1927, given this turbulent, emotional (even embarrassing) situation, Amy packed her bags and caught a train to London in search of a new life. She was in another of her dread-and-fled escape modes. But, no matter where she ran, she could never escape from herself or her old habits. For Amy, the streets of London were not paved with gold and nor did the capital change her luck in her prolonged love affair.

Amy first met her boyfriend through a family connection when she was still a Boulevard pupil back in 1921. She sometimes visited her Aunt Evelyn at Bridlington. Her son often played tennis with a former Swiss Army Officer called Johann Arregger – or Hans for short. Hans worked in Hull for the Swiss Consul and travelled around the UK, but initially lodged in Bridlington. Amy was immediately attracted to him with his sexy European accent. This was before she left for university in the October of that year. Understandably, her parents were anxious that their daughter (then 18) had a crush on a 27-year old foreigner – and even more annoyed that he was Roman Catholic.

Perhaps from Amy's idiosyncratic perspective, she might have thought, there is nothing wrong with falling for an older foreigner. After all, her (paternal) grandmother had not only fallen in love with an older foreigner, but also married him by the time she was 16-years old. And look at how successful their union had been as they approached their Golden Wedding Anniversary (5 January 1924).

Amy was useful to Hans in that he could practice his weak English upon her – spoken and written. Young, naïve Amy, however, always hoped for more out of their relationship; put simply, she wanted to become his wife! Generally, it was a one-way street; he played with her emotions and strung her along. He often joked he was not the marrying type – well, not to Amy as it happened.

Even from the earliest days of their relationship, Hans "took a cruel delight in teasing her about other women in his life" to make her jealous[74]. This is obvious in one of her letters soon after she became an undergraduate at Sheffield. On 15 December 1922, she asked him, "How do you like being at No.151 Boulevard? I expect you will be quite settled down there by now. Are you still as enamoured of your all-round sporting land-lady as you were?"[75]

After two years of trying, Hans eventually persuaded Amy to become his lover. This was in Sheffield during 1924 when she was a student away from home[76]. She reluctantly agreed, in the false hope that it would hasten their wedding day. In a vulnerable and irrational state, she frequently dropped hints that they must get married. Uncertain Amy even brought her superstitious beliefs to bear upon the matter. She wrote to him twice about different taboo rituals and strongly implied that he must propose. In one letter, she told him, "I slept with a piece of wedding cake under my pillow so that I would dream about my future husband"[77]. Around one Halloween, Amy tells him that if a woman looked in a mirror, she would see her future husband"[78]. He, obviously, never took any notice of these 'future husband' hints and omens.

When individuals are continually having their confidence undermined, they sometimes turn to superstition for strength and support. As a historian who has written extensively about taboos, I am keen to highlight the pronounced superstitious side of Amy's personality. Because of her reliance upon these folklore beliefs, various people gave her Good Luck charms over the years. When she first moved to live in London, for example, she displayed a lucky black cat ornament from Hans in her dismal hostel[79]. In Amy's own biographical chapter, she used the words 'luck', 'lucky' or 'unlucky' six times[80].

Their love affair continued after she moved to London. Amy struggled to find work and the jobs she found produced a meagre income. Her £50 loan had come solely from Hans, and she was determined to repay him every penny. It was an uphill struggle, but when she settled the final instalment, she handed in her notice at the dreary job and began to look for other employment. Long after they became lovers, Hans continued to taunt her about other women. So when he revealed he was engaged, she initially dismissed it as yet another example of him toying with her emotions. Ignorance is bliss.

Image 25: Amy was very fond of lucky black cat pictures and objects. Only in Britain are black cats lucky - especially aboard a ship - elsewhere, white cats are the preferred good omen. Hessle Road trawlermen were often keen to have a cat on board because of their nine lives. This Lucky Black Cat greetings card from Hull was not sent by Hans to Amy, but I could not resist inserting it here. Courtesy Keith Parker.

Amy's inability to get married to Hans was exacerbated in the autumn of 1926 with the news from Hull about her sister Irene. She became engaged to marry her loyal boyfriend Teddy Pocock, a cheerful, easy-going chap. Ciss and Will were over the moon – Ted was the nearest they had come to having a perfect son. He had a safe and steady job with the powerful Shell-Mex Oil Company; played piano when Irene sang at many parties; and he could handle Irene's wide and wild mood swings. Much to the anger of her parents, Amy did not attend her sister's wedding, which took place on Saturday 28 May 1927. She could not afford to travel up from London. Amy was not at all happy when she learned that her father gave the happy couple £200 to buy a brand new house in Lake Drive, near Hull's East Park. In some ways, Amy must have felt like the 'black sheep' of the Johnson family.

In sad and stark contrast, Amy began to sense that Hans was definitely two-timing her from the late summer of 1927 onwards. During one of his UK business trips, he met Connie Richards in her home town of Liverpool. She was the same age as Amy, blonde, and a Physics graduate with a Master's degree - far more academically qualified compared to Amy. Furthermore, she had a very good job working for the BBC (British Broadcasting Corporation) in its early days.

Meanwhile, Amy's prospects were still bleak. She felt, "London completely took my breath away... [it was] noisy and expensive and I felt lost and lonely. I knew no one...So far as finding lodgings was concerned, history repeated itself. Restless and dissatisfied, I changed continually, living in women's hostels when funds were low and in furnished bed-sits when jobs materialised"[81].

History also repeated itself because, for a long time, she never found any satisfactory or well-paid employment in London. Then "my luck in getting a job was definitely the turn of the tide". This was not thanks to her efforts, but to family connections. A distant cousin in London introduced her to "a solicitor who had known me as a small child and knew my family well. Taking pity on me, he offered me work in his office, beginning at £3 a week"[82]. Her job was at Crocker's, the prestigious London solicitors. I strongly suspect that her father Will wrote a few letters to family members in London to help his daughter, without her being fully aware that he had pulled a few strings – it was not as lucky as she believed.

Amy was upset when Hans mentioned he was engaged to marry a British woman; but she was devastated to learn that he had actually got married on 14 July 1928 at St.Patrick's Roman Catholic Church in Hull. Within six months, his wife Connie gave birth to their son. Amy was still bitter about Hans' two-timing – even ten years later[83]. Betrayed.

Her world shattered. She was humiliated, depressed, embarrassed, lost face in front of her friends and family, and probably felt that many people were laughing at her behind her back. She was a failure. Her head was in a spin. Would she have a nervous breakdown again? She must escape, run away from it all. Eleven years earlier, after the cricket ball smashed into her mouth, she took to the unladylike bike. This time, her mechanical means of escape was the unladylike aeroplane.

(9) JOHNNIE: One of the Lads

"The mere longing to fly was not new" Amy stated, "I had always, subconsciously, wanted freedom and adventure and I must have felt flying could give these to me"[84]. Like the bike, flying was another form of fleeing into freedom. Never face up to problems; just run away and, with a bit of luck, the problems will go away. Never face disgrace, just escape the place.

Outdoors, in the open air, Amy found freedom from people; freedom from pressure; freedom from problems. All alone, there was adventure without the cumbersome consideration of what others might think, feel or do. Up high in the clouds, she found her ideal escape from everyone.

After some long delays and frustrations, she took her first-ever flying lesson on 15 September 1928 – barely two months after Hans' wedding day. Even before his wedding, on a downward spiral of despair, she had happened upon and joined the Clubhouse at the Stag Lane Airfield, Edgeware in North London – not too far from the one-room flat in Maida Vale she shared with her loyal friend Winifred Irving.

Aeroplanes were the latest craze for the snobbish upper classes with money to burn. Amy was merely an office clerk on a lowly income and struggled to pay for one lesson at a time. She did not only have financial barriers to overcome, but also powerful sexist attitudes against female pilots in general. Aviation was virtually a 100% man's world in the 1920s. Everyone knew that; but for tomboy Amy, that was an added attraction as to why she must enter this male bastion. The more men tried to put her off, the more determined she became.

Amy did not want to be just another woman flyer. Some aristocratic women were doing just that for a mere jolly. She wanted to get to know the workings of the engine behind the propeller. This meant she had to bore deeper into the masculine world. It so happened that the de Havilland Aircraft Engineering workshops were located at the Stag Lane hangars. Within the very male-dominated world of aviation, it was even more macho amongst the mechanics. Yet Amy was up to the challenge. She thoroughly enjoyed working on engines, alongside the lads, in her oil-stained overalls.

Chief mechanic Jack Humphreys, fortunately, took a shine to her. He spotted her determination and potential, so provided her with some sort of shield from the mechanics' sexism, foul language and banal banter. After weeks of resistance to her trespassing into the world of engineering, the workshop lads gave her the affectionate nickname of 'Johnnie'. Obviously, a joking pun on her surname, but perhaps the blokes also detected Amy's tomboy side and encouraged it in her. By masculinising her name, they gave her a convoluted stamp of approval into their male territory. She loved it. Indeed, 'Johnnie' became a badge of honour when she became famous – 'Call Me Johnnie', she proclaimed[85].

Amy dovetailed her time in the workshop with learning to fly. Because her finances were low, her progress was slow. She also suffered a degree of resentment from some of her flying instructors – men, of course. Yet she persevered, paying for one lesson at a time; and saving up each week for the next. Lessons cost 30/- per hour (£1.50 = £60 in 2015). She often went without food and her hermit qualities came in useful once more.

I believe she was largely successful in gaining acceptance into the masculine arena thanks to her experience at the Boulevard School. Elizabeth Grey (1966) outlined how Amy found boys "worthy opponents not only in work, but also in play. Their games were far more fun than girls' games...the boys liked her. She was a 'good sport'...there was no lack of competition to walk arm-in-arm with her through the [West] park when school was over – though it was strictly against school rules" (p4) – not that school rules ever deterred Amy from doing whatever she wanted to do.

No doubt, some of those arm-in-arm boys were sons of top skippers or chief engineers. Amy, intrigued by community tales of perilous Arctic trawling trips, not only heard and absorbed much of the fishing folklore, but also tuned into the Hessle Road lads' sense of humour and a range of their swear words. I would suggest, therefore, that her adolescent years at school were a tremendous 'education' in socially interacting with working-class lads.

Johnnie would not find it too difficult to become 'one of the boys' thanks to her 'apprenticeship' at the Boulevard. Her early experience enabled her to chat and joke on equal terms with the Stag Lane blokes and she knew how to give as good as she got in the workshop banter. She felt at home in their company. Amy was happy. Johnnie was in her element.

Another payoff from her Hessle Road childhood was her love of unladylike subjects at school such as algebra, geometry and trigonometry. Mathematics and engineering go hand in glove – one goes with the other. The two suddenly fitted together perfectly. She already had the mental makeup to marry maths and machinery into one whole. Click.

When she bravely entered the world of engineering as a trainee apprentice, she would definitely have been joked about behind her back – the way guys do in a group, and especially when a woman was on her own. Yet she coped with this sexist ridicule and overcame a dreaded embarrassment in her life. Being laughed at was obviously a minus for Amy at that stage. However, life is a balance between pluses and minuses. The hidden plus for Amy was because engines were 'un-feminine'. This inner, deeper motivation enabled Johnnie to overcome the Amy impulse to run away from a threatening situation. In essence, working with inanimate machines was a buzz - in more ways than one.

July 1929 was a time of good and bad news for Amy – but even the black clouds concealed a silver lining. The happy news was that, on 6 July, she gained her Pilot's "A" Licence. This boosted her self-esteem; the licence was an essential step in her long-term escape plan. Amy had just had her 26th birthday on the 1st July, so it was a wonderful belated present.

The extremely bad news was that Amy's sister Irene suddenly, and unexpectedly, committed suicide. Happily-married Irene Pocock – the most beautiful, liveliest and successful of the four Johnson sisters – killed herself on Saturday 27 July 1929. Her husband Teddy had left early that morning for a whole-day business trip. Irene seemed extremely happy when he kissed her farewell. Her younger sister Molly called by during the day and saw no problems on the surface. Yet when Irene had the house to herself, she calmly wrote a suicide note. She placed it neatly on the living-room table, then stepped into her kitchen, closed the doors and sealed all the windows. She carefully positioned a cushion on the floor in front of the open oven door. All the gas jets were turned to full. She lay down quietly, rested her head and began to take in deep breaths. Slowly the room filled with the fumes of coal gas – carbon monoxide poisoning, the silent killer. That evening, Irene's body was found by her housemaid.

Everyone deals with death differently; but suicide takes grief into a deeper, darker dimension – especially when the self-slaughter was by a young person with so much potential and a close member of one's own family. Suicide raises profound questions about life and death. It was a time for Amy to question what she intended to do with her own life, especially when it could be cut short.

Amy's biographer Gillies saw Irene's death as a major turning point. Amy recalled her pre-school days when she and Irene were growing up in St.George's Road: "Of my little sister Irene, who most tragically died at the age of twenty-three, I have the softest and tenderest memories. Pretty and always very delicate, it seemed my job to protect her"[86]. Added to

Amy's string of recent disasters, she failed to 'protect' Irene in her time of need. Amy might also have felt pangs of remorse in that she had not travelled up from London to attend her sister's wedding. Had she put a jinx on the couples 'happy marriage'? Guilt.

Amy's parents and close supporters at Stag Lane - such as Jimmy Martin and Jack Humphreys - were extremely worried about Amy's state of mind. They knew she was a nervous creature and had recently had a breakdown – now this! They believed her sudden impulsive ambition to fly was a form of death wish, a disguised and elaborate method of suicide[87].

Image 26: The Johnson Family are here celebrating at Grosvenor House, London on 4 August 1930 after Amy's return to the UK from Australia. She is in the middle with Ciss and Will either side, plus her two remaining sisters Molly and Betty. Each of the five Johnsons must have felt the loss and absence of Irene in their own personal way. Courtesy Hull Daily Mail.

I only partly go along with this death-wish view. In essence, Amy was undaunted by death. Her darkest dread was people laughing at her behind her back. Given that, she fled. Now more than ever, she needed her mechanical means of escape. Both Hans' marriage and Irene's suicide doubly determined Amy to pursue her airborne dreams. She frantically needed to fly off alone beyond British shores to find fame, fortune and adventure. Where her father failed in the Rocky Mountains, she would succeed in the air. She did not seek gold in the hills; her goal was the silver lining of the clouds in the heavens above. She was the beautiful, red haired princess upon her horse with wings. In the distance below, she could hear her mother gently playing the piano. Nothing would stop Amy now. Yet there was one major factor slowing her down – the lack of money.

She worked upon her dad's sympathy as only a daughter can with her father. She was, after all, not just his tomboy, but his little girl too. Within forty days of the suicide, Amy had cajoled her dad to switch the financial support he had once given Irene over to fund her flying ambitions. "At first, Mr Johnson hesitated, but he knew Amy was determined to become a professional pilot. Finally, he agreed she could leave Crocker's and gave her a small weekly allowance to live on for the next six months. He also agreed to pay for her training"[88].

And so it happened. The cash began to flow from Hull to London into her account. During September 1929, Amy eagerly quit her job at the London law firm. Now she could spend all her time at the Stag Lane Aerodrome working on her engineering and flying qualifications.

On 10 December 1929, Amy gained her Air Ministry Ground Engineer "C" Licence in aircraft maintenance. It was a momentous achievement – historic, indeed. She was the first-ever woman in the world to qualify as a ground engineer. She was riding high. Added to that, the 1920s/30s was a time of record-breaking flights around the globe. The achievement of Australia's Bert Hinkler was of particular interest to Amy. In February 1928, he flew solo from England to his native land in less than 16 days – thus winning a £10,000 prize. His record still stood – unbroken. Only a handful of (wealthy) women were attempting and breaking such records. Amy shared her secret dream of beating Hinkler's solo flight with two close friends at the Stag Lane hangars (flying instructor Captain James Baker and Jack Humphreys). In her mind's eye, this was her grand escape plan. The Phoenix was rising from her ashes.

Fate lent a hand on 8 January 1930 when a London Evening News journalist visited the Stag Lane Clubhouse and happened to hear about Amy's desire to fly to Australia. Smelling a good news story, he immediately conducted an interview with her. No woman had ever attempted such a daring venture. A tidal wave of national publicity followed. Nevertheless, fame did not bring fortune for her flight. She wrote countless begging letters to a variety of big businesses seeking sponsorship. Not one sent a penny of support. Things were looking bad.

It was not a good time to obtain financial backing. Just seven months earlier, the devastating Wall Street Crash had happened on Black Thursday, 24 October 1929. It was the start of the Great Depression: banks closed; markets collapsed; businesses bankrupted; factories shut;

and millions on the dole. After Amy had declared to the world that she planned to fly solo to Australia, she was committed. She had put her head on the block. Her mind might have flashed back to her Boulevard days when she proposed her Panama hat revolt. Then, she had stuck her neck out, put her plan into effect, was abandoned, and none of her friends supported her. Her big idea came to nought. She paid the price and took the punishment and humiliation alone. She never forgot that isolation. Amy "was now confronted with the choice of risking her life or becoming a laughing stock"[89]. It was do or die.

The 10 March 1930 was a particularly good day. She obtained her "A" Licence as a Ground Engineer. Even better, after much persistence, her fish merchant dad agreed to pay £550 so that Amy could buy a two-year old aircraft she had set her heart upon. By today's (2015) equivalent conversion, that would equal £30,000. This grand sum was on top of Amy's everyday living expenses and training costs that AJK had been paying during the previous seven months. In a convoluted way, Hessle Roaders sponsored Amy's one-way ticket to Australia. I doubt very much that the Cod Farm lasses got many, or any, increases to their basic pay during The Hungry Thirties. The national mantra was "Austerity, austerity, austerity". Times were tough. It was a period of severe poverty for many; but financial reality did not buffet Will's daughter - cocooned within a privileged bubble.

Author David Luff clearly stated "Amy's flying career could never have succeeded were it not for their [AJK's business] financial backing"[90]. A few weeks before Amy planned to fly, her only sponsor was her dad. Amy would never have made her solo flight without the fish dock profits. She would barely have been a footnote in aviation history. There would have been neither Jason nor Amy Johnson in the city's heritage.

Although Amy was wary about her family's ownership of the Cod Farm, she was not so averse to taking sponsorship money from the AJK business in order to pursue her own expensive adventure. Perhaps her father recalled that he too had been 'subsidized' by the low-paid, casual Hessle Road labourers when he trekked off, starry eyed, to pursue the Klondyke Gold Rush fantasy 32 years earlier (1898).

Even though her parents were providing financial support, they were worried about Amy's health. "The strain of taking examinations and the stress of planning the flight to Australia had sapped Amy's strength. It would be suicide to fly...Slow down advised her parents, recover your health and try later...Stubborn as ever, Amy would not listen"[91].

One month after her father's donation, Amy was given an even bigger boost. Lord Wakefield, the Castrol Oil magnate, finally agreed to give her financial support on 16 April – plus the promise to refuel Jason en route to Australia.

(10) LUCKY JASON: Have Plane, Will Escape

Exactly one week later, on 23 April 1930 (St.George's Day), Amy finalised the deal to purchase a Gipsy Moth bi-plane – a mere 12 days before she actually took off on her ambitious journey. She moved fast and achieved much in a short time; did well and deserved full admiration. She also made a tiny, tiny acknowledgement to the corporate sponsorship from the AJK Empire. Two key aspects of Amy's aeroplane had strong, yet very secretive, links with her Hessle Road roots. They were in its colour and name. Both subtle aspects warrant reflective examination. Amy did not choose either lightly.

COLOUR: Amy's second-hand plane was originally dark red (maroon). She was obviously on a tight timescale and limited budget, yet she added time and cost to change the colour of the whole plane. For no logical reason, she ordered that the fuselage be re-painted bottle green (with silver touches). The repaint was described as an extravagance "especially as an extra coating would make the plane heavier...Any frivolous ounces made the plane harder to manoeuvre, particularly at take-off when it was carrying the most fuel"[92]. Some of the Stag Lane pilots and ground staff must have thought Amy was crazy. So why on earth did she make this unnecessary and fanciful change?

We will never know and it is open to speculation – so I will. As we saw with Amy's headgear at her Boulevard School, she relished breaking the rules. I believe she was motivated to do the same with regard to a powerful taboo within Hull's fishing community. The most dreaded, ominous colour was green.

There were two superstitious expressions on Hessle Road: "If a girl had a green dress, she would never wear it out" and "Green is followed by black" (meaning widow's weeds – black funeral garb). It was an ill omen if a house had green curtains, wallpaper, furniture, ornaments, or paintwork - especially on its front door. There are not many green cars in Hull showrooms because they take ages to sell. Auto dealers are happy to order a green car for a customer, but rarely stock one or display it on their forecourt.

Image 27: Amy's green machine. Jason is proudly displayed at the London Science Museum. It was built by de Havilland in 1928 and known as the DH60 Gipsy Moth. Some of the specifics are: open cockpit; bi-plane; fabric-covered wings; 30-foot wing span; 100 horse-power engine; registration G-AAAH.
Jason was only the twelfth of around 1000 Gipsy Moths built at Stag Lane near Edgeware - where Amy worked and qualified as an engineer. So she would have known her Jason inside out before she set off. Courtesy Hull Daily Mail

Green was very unlucky on Hull trawlers. If a young deckie-learner or trawlerlad came aboard wearing a green jumper or scarf, he was soon told off and the garment tossed overboard. I list many such examples in another book - and probe theories (Green Man links etc) about the origins of this profound and unusual colour taboo – so I will not repeat them here[93]. Amy, controversial as ever, declared to her father, "green is my lucky colour"[94]. She even took this one-step further and had a special flying suit made in matching bottle green material[95].

There are, of course, superstitious parallels for this type of contrary behaviour. Contradiction and superstition go hand-in-hand. Take, for example, the unlucky Number 13. Bingo callers often shout out "Lucky for some" and the players call back "No.13". Therefore, some people embrace the No.13 as their lucky number. A World War Two aircrew certainly took this 'Tempting Fate' approach when they deliberately named their Halifax Bomber 'Friday the Thirteenth'. It survived many raids and the war - then the RAF scrapped it[96].

It seems that Amy adopted the Celtic stance of wrapping herself in green – the colour of the little people (leprechauns) in the Emerald Isles, hopeful of attracting the 'Luck of the Irish'. She was a rebel who liked to break the rules now and again. She knowingly went in the opposite direction to the Hessle Road people - perhaps to distance herself from her fishing roots.

Superstitions can sometimes be twisted around to produce the opposite meaning. For example, the anti-female ditty: "A Whistling Woman and a Crowing Hen bring the Devil out of his Den" has been reversed into: "A Whistling Woman and Crowing Hen / May bring the Devil out of his Den / But the Woman who whistles / And the Hen that Doth Crow / Go far in Life wherever they go"[97]. This last line illustrated what Amy did all her life. She embraced the negative attitude towards green and transformed it into her lucky colour.

Having speculated about a counter-superstitious reason for painting her Gipsy Moth green, Amy had a direct reason too - she was against the colour maroon. In her autobiographical account, *Myself When Young*, she recalled that her Boulevard "school uniform was maroon, a colour at that time I heartily disliked...Not only did I have to wear this hideous colour, but in addition had to have my long curly hair in two plaits with a maroon bow at the end of each"[98]. The net result was that instead of her plane being the hated maroon, she had it re-painted green – the colour hated by many Hessle Roaders. After all, she could have chosen blue, yellow, cream or whatever, but she homed in specifically on green (Images 27 and 28).

NAME: Amy discreetly acknowledged the enormous debt she owed to her dad's fish dock firm when she made the vital choice of a name for her very first aircraft.

The naming of Jason was like a very, very private Johnson Family joke – known only to the intimate few. When the name was publicised, the media followed the line of thought echoed by one of Amy's earliest biographers, Herbert Banner. He took the bait as intended and proposed the view that Jason came directly from ancient Greek mythology – the legendary Jason and the Argonauts. The world at large thought it was "a fitting name, since this Jason was to fare forth in search of the Golden Fleece of a mighty achievement for England's honour"[99]. The 'Golden Fleece' was obviously a link with Australia's massive sheep-rearing activities.

In reality, however, Amy's Jason had NOTHING at all to do with Classical Greek adventures. There was instead a covert link with her family and their fish business. Amy wrote to her father, "I don't think anyone connects it [Jason] with kippers"[100]. This quote comes from Constance Babington Smith (1967) and implies that JASON was the brand name of a Hull Kipper marketed by AJK. Try as I may, however, I have not yet been able to confirm this dimension of the Jason story. I wish I could find a label for the 'Jason Kipper' to reproduce here – no luck so far, but I will continue to hunt (like Jason in search of his Golden Fleece!). When her father read her remark about the kipper, he replied, "Jason will bring you luck".

Image 28: Here is Amy with newly-painted Jason – getting ready for her flight. One of the key reasons she bought this particular aircraft was because the previous owner had fitted it with two extra fuel tanks. In total, her three tanks held 80 gallons. Cruising at 85 mph, in ideal conditions, meant she could be airborne for 13.5 hours at a stretch = almost 1150 miles per day (not that she ever did that in practice). This enabled Amy to fly longer than Bert Hinkler could in 1928. More time in the air meant less landings – Amy's weak point. Courtesy Hull Daily Mail.

Although I have failed to find a Jason Kipper label, there is definitely a strong Jason link with the Johnson family business. JASON was both the registered trademark and telegraphic address for AJK (Images 29 and 30). It may be that this name juggles with the 'Andrew' and 'Johnson' - swapping the 'A' for the 'ohn' to arrive at JASON. Alternatively, 'Johnson, Andrew, Knudtzon' can be shortened to JAZON, but looked better as JASON. Equally, from the AJK perspective, the Greek legend of Jason fitted perfectly with the naming of their trawler fleet that drew upon Classical characters (Image 6). Telegraphic addresses were deliberately short and evocative so that customers would easily remember where to cable their orders.

In a nutshell, Jason the plane obtained its name directly from AJK, and not a jot to do with Greek legends. Hull's St.Andrew's Fish Dock, not Athenian literature is the true source - not Hellenic, but rather 'Hullenic'!

For all the riches Amy received from her father's firm, the naming of her airplane was one small crumb of comfort in the way of payback. Many sponsorship deals insist upon a reciprocal arrangement - positive publicity in exchange for financial support. As will be seen after her return from Australia, Amy repaid the actual cash value (£550), but never a penny in kind or public acknowledgement to the fish dock business.

Knowing Amy's sense of humour, I suspect she had fun with changing the paintwork and naming her special aircraft. Finch (1989) tellingly wrote, "at times Amy's attitude to the Australian flight was almost childlike – as if it was only a glorified bicycle tour" (p23) – with her thermos flask and sandwiches for the outing.

Before going on to look at her world-famous flight in Jason, I would like to re-emphasise a point about Amy. No one whose powerful grandparents were deeply involved in the trawling industry could but be influenced by the fishing heritage of Hull. No one whose father owned and ran a massive fish merchant business could be unaffected by the fishing culture. I believe that the trawling heritage that surrounded Amy Johnson in the Hessle Road community would have permeated her outlook in many profound respects. Some of the best examples of this saturation were found in Amy's highly superstitious rituals.

Image 29: I was delighted when Janet Scott emailed me this AJK 1934 letterhead. The importance of the document lies in the telegraphic address of JASON (top left). It is the direct link between the name of Amy's famous plane and the debt she owed to Hull's fishing heritage. Indeed, it could be argued in a convoluted way that the £550 that bought Jason was funded out of the labour of the port's fishing families. Courtesy Janet Scott.

(11) AUSTRALIA: Fingers-crossed Landings

Just prior to her solo flight, Amy's mother Ciss sent parcels plus "a welter of good luck tokens and charms. Amy was surrounded by black cats and an elephant". She replied in a letter "I can't help but have a lucky trip"[101]. Despite being strong Methodists, the Johnson Family also pandered to Amy's fetish with superstitious charms. There is a generally accepted maritime view that "of all seafarers, there are none more superstitious than fishermen". Amy's superstitious personality certainly demonstrates how the Hessle Road Fishing Community pervaded the walls and beliefs of the Johnson household – and reached their eldest daughter.

Harada & Hunter (2013) outlined the ways superstitions can seep into a society: "As we go about our lives, we are unconsciously immersed within the cultural fabric of the society we live in. To a certain degree what we think and do is greatly influenced by the values, beliefs, and meanings adopted by that society. Embedded within our belief systems are a wide range of customs, rituals, taboos, and behavioural codes that have basis upon superstition" (p1)[102].

It is also possible that Amy deliberately adopted a superstitious outlook because it ran counter to her parents' Methodism. We have seen how Amy almost enjoyed being contrary to authority figures - be they parents or teachers. Furthermore, the illogical working-class folklore rituals were perhaps an additional draw because they too went against the grain of the middle-class values within her home.

Superstitions can sometimes be like a pick-n-mix sweet counter – people select some and reject others. With certain strong Hessle Road taboos, Amy deliberately went against them as we saw with the colour green. Such contrary, selective behaviour was not uncommon amongst fishing family members.

Remarkably, it was less than one year since her first-ever, single flight (9 June 1929), that Amy prepared Jason for the trip. The take-off took

Image 30: These two images clearly display the registered Trade Mark and Telegraphic address of "JASON" on the Fish Dock offices of AJK c1930. It is interesting to note that Andrew Johnson chose a name based upon a Greek legend. It certainly echoes the Classical theme of his trawler fleet with names like Achilles (H.109). Courtesy RAF Hendon Museum.

two attempts because the overloaded aircraft had extra fuel and spare parts. Twenty-six year old Amy flew from Croydon Aerodrome on 5 May 1930 (07:45 hours) heading southeast for Australia. There was only a tiny crowd to see her fly away. Amongst them, of course, was her anxious father Will - probably still grieving and coming to terms with the suicide of his favourite daughter Irene only ten months earlier. His heart must have flown with Amy as he watched his first-born child soar out of sight. She was following in his footsteps – and he was proud of her sassy spirit. He may have felt: You show 'em girl!

The scale of her venture was enormous. Previously, Amy had only been on short holidays to the Alps (to stay with Hans and his family), never mind a journey of around 10,000 miles covering alien terrain. Her planned route followed as straight a line as possible with twelve scheduled overnight stops: Vienna; Constantinople; Baghdad; Bandar Abbas; Karachi; Allahabad; Calcutta; Bangkok; Singapore; Sourabaya; Atamboea; and Port Darwin[103]. Britain, having its Empire at the time, was a great help in her planning this route.

She had even made definite plans for her return journey to Britain in *Jason*. As mentioned earlier, I do not intend to describe her flight in any great detail, but would urge any reader interested in the odyssey of a solo woman's battle against the elements to read her life story.

Nevertheless, Amy's daredevil venture does demand some attention as she progressed 'down under'. Her first stopover in Vienna illustrated two interesting (perhaps disturbing) aspects of her flying ability and personality. She described her perfect landing: "I was still sufficiently a novice...to regard a really good landing as a piece of luck"[104]. This was quite a confession by a qualified pilot. Funny though it may seem, Amy never got the hang of landing any aircraft with great confidence! Such open honesty testifies to her bravery, foolhardiness and her dependence upon luck.

At Stag Lane Aerodrome, a number of experienced pilots were extremely sceptical that greenhorn Amy would get very far on her 'hair-brain' journey. Many felt that she "was not a born pilot or natural flyer ...she never quite mastered the knack of the perfect landing...Every time she took to the air she knew she'd have to face the anxiety as to what would happen when she tried to land"[105]. The general view was that Amy was a better engineer in the workshop than a pilot at the controls about to land.

Her Vienna landing reinforced a second point that I am keen to emphasise: her "I-cannot-stand-being-laughed-at" personality trait. Although Amy found the Austrian ground engineers friendly, their 'aggressive efficiency' was distressing. Exasperated Amy declared, "the mechanics absolutely refused to let me do anything on my engine...they endeavoured to amuse me by a story of their last lady pilot visitor...who also insisted on looking over the engine herself. Now, if there's one thing I cannot stand, it's the idea of being laughed at behind my back - so I left them to it"[106].

Within her open cockpit, Amy had only very basic instruments to guide her: the standard P-compass, altimeter, airspeed indicator, turn-and-bank device, and perhaps a fuel gauge. Surprisingly, she had no radio or lights. In her 'village shop' she had a few essentials to help her survive in an emergency: thermos flask (and sandwiches), first-aid kit, mosquito netting, cooking stove, flints, air cushion, tools, tyres, a spare propeller (strapped to the side of the fuselage), and a revolver. From Amy the Hermit's perspective, she did not mind doing without the luxuries of life for a certain time, nor being alone.

After overnight stays in Constantinople (modern-day Istanbul, Turkey) and Aleppo (Syria), Amy headed for Iraq, where she encountered a sandstorm of terrible ferocity. Amy's personal account on 8 May recorded: "Suddenly Jason and I dropped a couple of thousand feet. The drop was so sudden and far that the propeller stopped...The machine dropped again...to within a few feet of the ground...helplessly blown hither and thither...sand and dust covered my goggles – never been so frightened in my life"[107].

Midge Gillies then described how, on the sands of the Iraqi desert, Amy "felt Jason's wheels touch down at around 110 miles per hour and tried to steer into the wind in an effort to slow the machine, fearing that the plane would at any minute flip over or hit some obstacle. When Jason finally stopped, Amy switched off the engine and struggled out of the cockpit...She pulled the machine round so that it faced into the wind... Her priority was to protect Jason, her only means of escape from the desert...trying to cover the engine...from sand and dust...It took her thirty minutes to lash the cover down...

"Having made her plane as safe as possible, she sat with her back to the wind on Jason's tail in a bid to fix him to the ground...she had no idea where she was...After three hours the wind began to die down and the air cleared...Jason's engine started at the first swing of the propeller and Amy pointed the engine in an easterly direction, hopeful that this would take her towards Baghdad"[108] – it did! This was but one of the many obstacles that Amy overcame during her trip – thanks to her guts and determination.

On 10 May, when Amy landed in Karachi, she was two full days ahead of Hinkler's world record from England to India. The world media now latched on to Amy's story and keenly followed every inch of her journey from then on. During her next two landings, Jason was damaged twice - especially north of Rangoon (Insein, Burma) where events conspired against her. Her engineering and makeshift skills were stretched to the limit, but she proved her metal.

It was the 15 May before Johnnie and Jason were able to get on their way again. Newspaper readers around the Commonwealth and beyond were gripped by her undertaking. Everyone knew that Amy now had no chance of breaking the solo-flight record to Australia. Yet that did not matter, the world was cheering brave Amy and tiny Jason all the way – willing them on to make it.

Amy only had rolled up maps of the rugged territory below her – some areas were unchartered. She navigated by looking over the side of her cockpit trying to recognise the landscape far below – if it was clear weather. No doubt, she recalled Headmistress Ada Knowles' geography lessons as she peered in search of landmarks such as the River Euphrates, railway tracks or various coastlines of the Bay of Bengal. Navigation was not easy in those early flying days.

Another difficulty was that Amy lacked any reliable weather forecasts and encountered a welter of "beastly" conditions. She never knew what she would have to face from hour to hour: hurricanes, fog, monsoon rains, intense heat, sandstorms, dense clouds, high mountains, cold, or endless seas (with no islands for guidance). Amy's bravery was beyond question and the ordinary people loved her for that. Some newspapers described her as 'the flying typist'. There were some in authority, however, who saw her in a completely negative light.

Image 31: This is Amy at Surabaya (Dutch East Indies) where the British Consulate was none too happy with her suddenly landing in his neck of the woods 20-21 May. He sent a negative message to the Foreign Office in London criticising her "sex, youth and comparative inexperience". Who knows, perhaps she interrupted his golfing plans! Courtesy RAF Hendon Museum.

The National Archives released volumes of official papers to mark the 75th anniversary of the flight. An Air Ministry document was a pessimistic memo, drafted 10 April 1930, before she left Croydon and focused on her lack of preparation. Worst of all, Mr. J. Drummond Hogg, the British Consulate in Surabaya (then Dutch East Indies, now Indonesia – East Java), condemned her venture by stating, "Some restrictions should be placed upon such foolhardy enterprises" in view of the "sex, youth and comparative inexperience of the pilot...Adventuresses like Miss Johnson should be protected from themselves and not permitted to venture alone upon such dangerous undertakings for purely selfish ends"[109]. Officialdom took a dim view of Miss Johnson's antics.

After departing Surabaya, she directed Jason toward Atamboea in Dutch West Timor, flying over a string of Indonesian islands. This was to be her penultimate stop before Australia; but it was no time for complacency. There were more unexpected problems in store.

Dusk fell suddenly. Low on fuel, she urgently needed to find the Atamboea airfield. Peering over the side, it was not in sight. Below, unseen, airport officials frantically jumped up and down and waved at Jason. What they could not tell her was that, a few days earlier, there had been a wild bush fire and the whole area was a blackened mess. She could see nothing of the landing strip from the air. Amy flew on over the thick jungle and deeper into the growing darkness.

She had to fly lower and lower. She had to land anywhere. Was that a field ahead? No choice - must land. It was not a clearing. Six-foot high anthills suddenly appeared everywhere. Amy veered Jason this way and that. Deadly stumps whizzed passed on the left, on the right. After an eternity, she brought her fragile craft to a halt. One wing rested on a mound. The gods were with her. Jason was OK. She had landed.

She switched off the engine. Immediately, a new danger threatened. Horrendous shrieks and loud yells filled the air. Instinctively, she reached for her gun. A horde of Timorese tribesmen rushed from their mud huts and surrounded her plane. They waved spears and big machetes above their heads. Some carried knives in their red-stained teeth. Wild hair waved in the wind. Amy gripped the revolver in her hand. What should she do? This was it. Her worst nightmare was a reality.

A tribal leader stepped forward. He touched Jason's green fuselage. He looked Amy in the eye and gave a gesture, a salute. He ranted. The crowd chanted. She could not understand any utterance. Then she detected the word "pastor". He pointed into the jungle. She lowered her concealed weapon – tucked it into her jacket and stepped down from Jason. He offered his hand. She took hold.

Although exhausted, Amy could not rest yet. She hiked miles through the tropical forest. Eventually, the group came upon a large log cabin - a sign of civilization, a sign of safety. Amy slumped onto the wooden steps and immediately fell into a deep sleep.

The next thing she recalled was a white man with a large beard shaking her by the shoulder. He was a Dutch pastor. He had rummaged through what he could from his minuscule rations for his unexpected guest. They communicated in what limited French each could muster, and enjoyed some cheese and a bottle of wine together. The loud blasting of a car horn outside the cabin suddenly interrupted their conversation.

The Portuguese commander of the elusive airfield had come to find Amy. Thankfully, he spoke good English and explained about the bush fire. The kind pastor waved Amy off as she disappeared toward Atamboea.

There, she slept properly, even though she was worried about Jason alone in the jungle. As she slumbered, the Daily Express banner headline in London read "MISS AMY JOHNSON – NO NEWS". Even no news was big news about Amy! This stemmed from an earlier report after the Atamboea ground staff reported she had flown over their airfield and vanished.

Next morning, a donkey was loaded with two canisters of aviation fuel. Then Amy and a party of airport staff trekked off and found Jason about sixteen miles away at Hilula. She spent two hours carefully filtering every drop of murky petroleum into Jason's main tank. It was enough for her ten-minute flight to Atamboea - the most primeval 'airport' she had ever seen.

There was no hangar for Jason and no shade for Amy in the baking sun. Charred wood and black dust were everywhere. The air was still thick with pollution. Breathing was difficult. Amy's eyes were sore with irritation and she could not stop coughing or clear her throat properly. The fuel for Jason was filthy with not only ash dust, but also red rust. Meticulous Amy (another of her father's inherited traits) painstakingly strained the fuel. One piece of grit could stall Jason's engine in mid-air over the terrifying Timor Sea. That would be the end of everything.

Officials, however, were wary. They had just had a devastating bushfire. Here was Amy handling inflammable liquid in the mid-day sun. The risk of another blaze was high. A guard stood watch as she worked on the tedious task. She toiled on, knowing she would have to spend another night at Atamboea. She could not sleep for excitement knowing Australia awaited the next day.

As Jason took off, a black dust bowl whipped into the air. Amy was free and on her way again. At last, she breathed fresh air deep into her lungs. Ahead was no picnic, for Amy had never faced such an endless stretch of water. So far, there had always been land within a few hundred miles. The Timor Sea was open water for 485 miles. There were demons below the waves. All pilots feared the deadly sharks – they hunted in packs. Yet the greatest enemy was within. Boredom and lapses of concentration could easily flood in. She had not really slept properly since she left England.

A couple of hours into the monotonous flight, Amy spotted menacing dark clouds ahead. Her usual strategy was to take a wide diversion around them. She did this and then got back on track. Her heart took a leap of delight. There on the distant horizon she saw a beautiful sight - a column of thick black smoke. It was a pre-arranged signal; a halfway house; a beacon of hope. Shell had ordered their 4120-ton oil tanker Phorus to anchor midway between Timor and Australia - three hours flying time. The crewmen jumped for joy, cheered and waved as plucky Jason made a low swooping salute over their decks. She tossed a piece of cake down to the men. The wireless operator hurried to the bridge and signalled to the waiting world that Amy was safe and on her way.

Escorted by two waiting planes, she landed in Port Darwin on Empire Day, 24 May - Day 20 of her odyssey. Australians adored Amy. During her first speech, in response to the welcome by the Mayor of Darwin, she said "Please don't call me Miss Johnson – just plain Johnnie will do" – and the crowd cheered[110]. The Aussie media nicknamed her Joystick Johnnie. She enjoyed the Australian sense of humour. "Amy chuckled at the caption beneath one cartoon which read: 'Our Amy has got more backbone than her father's kippers'[111]. When the news reached Hull of Amy's safe arrival, Will Johnson immediately sent "out a radio message to the fishing fleets in the North Sea, saying that Amy had the blood of the trawling industry in her veins"[112]. True.

She flew her Gipsy Moth from city to city on a national tour until, on 29 May, a serious crash landing in Brisbane brought her flying to an abrupt halt. Jason was wrecked, but Amy was unharmed. On 3 June, news reached down under that the King had made Amy a Commander of the Order of the British Empire (a CBE). After six exhaustive weeks of speeches, galas, and official events, she sailed from Australia aboard the P&O liner Naldera to Port Said in Egypt[113] (Image 32).

From there she flew as a VIP passenger with Imperial Airways back to Croydon on 4 August 1930 (a public holiday). A smart-looking Jason was waiting for her arrival at the aerodrome – having been repaired by Aussie engineers and shipped home to the UK (Image 33).

Image 32: After her exhausting solo flight and the adulation of the Australians (plus her crash landing at Brisbane), Amy was relieved to step aboard the large P&O Passenger Liner Naldera. She was, of course, their VIP guest. From poverty and obscurity, she had ascended into a life of luxury and superstardom. This was just a taste of what was to follow when she arrived back in her home country – and she relished it (initially, at least). Courtesy Clyde Built Ships Database website.

(12) FAMOUS and AIMLESS: Unlucky in Love

Britain went ballistic. About one million people lined Amy's 12-mile route from Croydon to London's Park Lane. There, she was given a grand reception in her honour at Grosvenor House (Image 26). This massive number of well-wishers was in stark contrast to the handful of people who saw her depart from the capital only three months earlier. Later, on 11 August, she flew Jason solo to Yorkshire where she landed at Hedon Aerodrome - east of Hull. Amy's proud grandfather Andrew Johnson, in his 78th year, was also there. He had just recovered from a stroke, but leapt up and down with excitement when he saw her[114].

Hull had its hero. Everyone wanted to see 'Amy, Wonderful Amy'. At this time, her parents still lived at No.85 Park Avenue. Thus, the Amy-Avenues link became firmly fixed in the public psyché. Hessle Road never got a look in – even within Hull.

One of my students told the story of being a Cod Farm worker when the world cheered Amy's success. Jack Anderson was a lad of fifteen in 1930 when all the AJK staff were called into the firm's biggest shed. All the Johnson Family managers were there, proudly announcing that each worker would receive a shiny half-crown coin (2/6d) to mark the important event (over £11 by today's value). Everyone was euphoric. Young Jack stepped forward to receive his piece of silver. He had never been so rich in all his life.

"But then," Jack ruefully added, "before we left, it was suddenly announced that a collection was being made to buy Amy a present of a car". Poor Jack felt obliged to drop his 2/6d into the large box manned near the exit. A week or so later, her father Will arrived at Cod Farm driving a brand new car. The bosses explained that Amy had about half a dozen cars given as gifts and so her dad was just looking after it for a while[115]. "The rich get richer and the poor..."

From the moment she became a superstar, a variety of gifts showered down on Amy from around the world. De Havilland gave her new aeroplanes. Various cars came her way, especially a brand new MG Mk 1 Salonette from Sir William Morris with a mascot of Jason on its radiator cap[116]. Many businesses, such as Dunlop, Waterman Pens, Gaumont, and Coco Chanel, were eager to have their products associated with her fame and success. Sponsors rarely act like charities; they expect something in return for their 'generosity'. Amy found herself committed, indeed trapped, by some of her sponsorship deals. Castrol (Lord Wakefield), the national Daily Mail, and Daily Express each extracted their 'pound of flesh' from her. She became ill through the pressure and stress.

Image 33: Here is Jason in all his glory after it had returned to the UK from Australia by sea. Emblazoned on the fuselage is the plane's name along with a white star. This was the symbol of de Havilland – with a gipsy moth image imposed upon it. Clare Bevan told me that the large white building in the background was the Croydon Aerodrome Hotel – the place where Amy slept the night before starting her famous flight on 5 May 1930. Courtesy Clare Bevan.

There was, however, one brand missing. It was her first and sole sponsor; the one that came to her rescue when the rest of the world shunned her appeals for money. They never received one second of publicity time from Amy. It was, of course, Andrew Johnson, Knudtzon & Company. None of the Johnson's fish products got promoted. Amy never held up a packet of the firm's boneless kippers – launched as a brand new initiative only two years earlier. It seems surprising that she did not once reciprocate her father and his fish dock business for all the financial sponsorship she received from them.

Oddly enough, Amy did not invite any member of the Johnson Family to the Buckingham Palace ceremony when she received the national gift of a CBE from King George V on 11 August 1930. Instead, she took along a couple of her Stag Lane mates. Why did she snub her parents? How did they feel about not attending this royal event? Why did Amy not want her Hull relatives mixing with her London set? Snubbed.

Amy was in big demand and went on a Daily Mail national lecture tour entitled "How Jason and I Flew to the Land of the Golden Fleece" during the Autumn of 1931. Obviously, Amy was keen to embrace the misleading Classical Greek imagery of her aircraft, rather than acknowledge the generous backing she received from the 'Jason' fish product brand. This is a clear example of the deception she created to deflect any attention from herself and Hull's Cod Farm.

It was during this period, however, she did accept an invitation to visit her old school at the Boulevard to unveil a plaque in her honour (25 January 1932)[117]. As she was in that 'neck of the woods', I just wonder if she called by her birthplace at No.154 St.George's Road or even drove along Hessle Road. I very much doubt she went anywhere near St.Andrew's Fish Dock to thank the workers personally for one of her new cars.

Image 34: There are around 77 workers in this staff photograph - the Johnson Family owners are in the middle at the front - in around 1908. I counted 47 Cod Farm lasses – over half the total. AJK were certainly a large employer on the fish dock and their Cod Farm business must have made a reasonably good profit over the years. Courtesy RAF Hendon Museum.

On 9 May 1932, after spending less than ten hours in his company, Amy accepted a proposal of marriage from Jim Mollison – the Flying Playboy. He was a terrible choice for Amy. Like Hans, Jim used her for his own selfish ends and enhancement of his public image. She was oblivious of men's motives. In some respects, Jim was partly Amy's ticket into London's high society life. They were to marry within three months of their engagement.

After her rapid rise to international fame, Amy became even more ashamed of her Yorkshire accent. As with many educated middle-class people - and encouraged by her mother - Amy adopted the Standard Received Pronunciation of the day (like BBC English). In addition, she continued to deny links with her hometown of Hull[118]. She was never an open ambassador for the city. Bob Finch pointed out that Amy "had no attachment to the place in which she was born and bred and, once she became famous, rarely returned to it. If anything the globe was more of a home than Hull"[119].

It would be wrong to suggest that Amy was alone in having a snooty attitude towards her Hessle Road background. Many people throughout Hull and the East Riding of Yorkshire often spoke of the fishing community in derogatory terms. I touched upon this social snobbery in my first printed book Village Within a City. I described when "the Hessle Road community entered the twentieth century, it was in many ways 'a village within a city': although a part of Hull, it was decidedly apart from it" (p6), and "The community's separateness was further increased by the attitude of other people in the city. Outsiders were heavily influenced by the narrow, but powerful image of 'drunken brawling fishermen'" (p51)[120].

I have made the case that Amy Johnson was ashamed of her Hessle Road roots and especially any hint of a link to Cod Farm. Gillies (2003) went much further and claimed that even her own parents embarrassed Amy. Her supportive evidence for this claim is that Amy did not invite them to her wedding to Jim Mollison on 29 July 1932: "she was ashamed of them and they were no longer important to her" (p215).

Ciss and Will Johnson were in a state of complete shock and anger when they found – just the evening before - that Amy was to be married the next morning at St.George's Church, Mayfair, London. Amy had previously told her parents to ignore newspaper rumours about any wedding[121]. They then lived in Bridlington and there were no more trains that night from the town. Despite not being invited, and a torrential rainstorm outside, they decided to bundle their girls Molly and Betty into the back of their car and drive through the early hours to be at the wedding. First, Will had to collect an overcoat from his office at St.Andrew's Fish Dock. It was around 3 am before they were able to continue on their two-hundred-and-fifty mile drive, along poor, winding roads [compared to today's motorways]. They were fuming at their daughter's gross disrespect. Was this the thanks for all they had done for selfish Amy? Was she ashamed to have them mix with her new high society friends?

Image 35: For many years, I showed this image of St.George's (Road) Wesleyan Methodist Women's Group in my 'Hessle Road Community Spirit' slide talk. Then one evening, someone in the audience from The Cottingham History Society pointed out that Amy Johnson was stood near the seated vicar - to our left. At a guess, the occasion was in January 1932 on a revisit to her childhood Sunday School – where she had been classed as a 'scholar'. Courtesy Fred Ingilby.

Despite their frantic journey, the four Johnsons arrived just too late for the ceremony. National newspaper reporters recognised her parents and informed them that the couple were already signing the register. They declined an invitation to go into the vestry and quietly remained aloof from the occasion. As the bride "walked down the aisle with her new husband, she failed to notice them. It was only at Grosvenor House [at the elaborate reception]...that reporters told Amy her parents were in London. She was horrified and asked someone to ring round the major hotels to try and find them, but the Johnsons were already on their way back to Bridlington"[122] – driving back up the Great North Road (A1) toward Hull. They had no wish to stay where they were not welcome. Hurt.

Amy's 'dread-and-fled' tendency was even more pronounced after she became world famous. There was no getting away from press photographers. Even as early as September 1930 she wrote a letter to top aircraft designer and friend Geoffrey de Havilland looking for sympathy. She told him "publicity would in time drive me insane and I'm therefore taking the cowardly action of running away from it...My one desire is to be left alone in peace to fly"[123]. Up in the air, away from everyone, alone in her cockpit, tomboy Johnnie could satisfy hermit Amy. She was happy to escape and be in her own company. She related better to machines than people. If they broke down, she could fix them. People were not so easy to put right. When it came to close relationships, Amy had various 'crash landings'. If she had love in her life - or thought she did - she turned her back on her parents and family.

Unlucky in love, Amy never found true happiness. She "was often irresistibly drawn towards the very men who did not shrink from exploiting her"[124] and "she frequently took up with somewhat caddish and elementary types of men who treated her rather cavalierly as that type sometimes does when they find a woman who is more enthusiastic"[125]. Generally, she was attracted to men who were indifferent toward her (as with Hans and then her husband Jim Mollison) and irritated by men who adored her (François Dupré, Jack Humphreys, and Peter Reiss[126]).

Amy was close at times to her father and dependent upon him for financial support and advice - both before and after she was famous. She fell out with her parents, especially her mother, when involved with men; but then leaned upon her family yet again after being hurt in love. A modern-day term, from her parents' perspective, was that she was 'high maintenance'.

Given Amy's unsettled romantic state, her emotional life oscillated between extremes. When in poverty and struggling to achieve fame through flying, she neglected her appearance – she wore greasy overalls in the workshop or, in the cockpit, she had petrol splashed all over her clothes and in her hair. On the other hand, when world famous and the money was gushing in, she was extravagant - spending upon dresses, pets, hotels, travel, her teeth and skin care. She seemed to live life lurching from one extreme to another. All or nothing.

In some respects, she mirrored the lifestyle of the 'three-day millionaires' who enriched her birthplace of the Hessle Road Fishing Community[127]. The crewmen's messy work on deck of gutting fish, going unwashed and unshaven for days on end was in stark contrast to when they came ashore, changed into their flamboyant smart suits, and had a spend, spend, spend mentality. Amy is highly unlikely to have mixed directly with Hull trawlermen in port; but she would have heard of their colourful suits, generous gestures, and heightened emotional lifestyle. After all, she did attend the Boulevard (mixed gender) school and children of skippers would have been in her class. Tales of the trawlermen's lavish lifestyle would have appealed to Amy's fertile imagination.

Amy was a good writer as can be seen from her work as an author and columnist for a couple of newspapers. Writing, in some respects, can be a form of escape and an isolating process at times. Her book Sky Roads of the World was very forward looking and positive about the future of aviation[128]. She was a highly intelligent and creative visionary – when she put her mind to it. She also kept a smart eye upon publicity and her public image. Indeed, the 'Amy look' – with her Coco Chanel garments and 1930s hairstyles - was copied by fashion-conscious women everywhere. She could have been successful in business, but rarely found the right financial partner; never anyone she was happy with at least.

As with Hans and her parents, husband Jim Mollison also provided Amy with good luck charms. Prior to her solo flight to Cape Town, South Africa on 14 November 1932, Jim gave her his St.Christopher token for a safe journey. All Amy's dependence upon lucky trinkets from her immediate circle strongly shows just how superstitious she was throughout her life. Like troops going into battle or astronauts blasting into space, Amy was laying her life on the line every time she took off - and especially with her patented fingers-crossed landing technique.

During her 1934 lecture tour, she told audiences why she could not give up dangerous flights: "the greatest gamble of all is the gamble with life and that is the fascination which attracts all who love adventure"[129]. To my mind, Amy had an affinity with the perilous occupation of Hull's swash-buckling Arctic trawlermen – 'last of the hunters'[130]. They too gambled with their lives every time they set sail to the distant fishing grounds. Like Amy, they too battled against the raw elements. A tremendous thrill was to confront a storm and survive – to come out alive. Finch (1989) summed it up as "the only true exhilaration comes when life is put in jeopardy and when the gruelling urgencies of a challenge are superseded by the euphoria of triumph" (p63).

She must have read or heard stories of Hull trawlermen surviving against the perilous odds. They faced numerous life-threatening dangers in the Arctic waters. For example, just a quick look at the port's trawlers between 1930 and 1934 reveals nineteen losses (thirteen wrecks, three sinkings, and three collisions) within only a five year period[131]. It is possible, therefore, that Amy identified with Hull trawlermen who also lived on the edge of existence – and who, incidentally, risked their lives to land fish for AJK's Cod Farm. I believe Amy intuitively absorbed from the Hessle Road fishing families much of what she abhorred: the folklore superstitious rituals, the live-for-the-moment three-day millionaires' lifestyle, and gambled with her life in a similar way to the Arctic trawlermen.

Image 36: Amy is here in one of her many cars. During the 1930s, she entered a number of car rallies – but never with any great success. During her time living 'in the fast lane', however, Amy had a couple of driving accidents. Luff (2002) related how, in April 1935, Amy was driving a Mercedes after a visit to her parent's home in Bridlington. She was "on the Fraisthorpe by-pass when an approaching motorcyclist pulled out from behind a car he was attempting to overtake. As he did so, he clipped the car's rear bumper and was thrown directly into the path of her car, with the result that his pillion passenger received fatal injuries. Although Amy was exonerated of all blame…it shook her confidence" (p.268). Courtesy RAF Hendon Museum.

Briefly, I would like to return to the title of this publication: 'AMY JOHNSON: Hessle Road Tomboy – born and bred, dread and fled'. Despite the fact that Amy was very much a Hessle Road girl, given her deep and long fishing family legacy, born and bred in the community itself, she very much dreaded any links to her father's fish dock business and the Cod Farm lasses – so ran away from any association with Hull. Amy was not true to herself or her proud fishing origins. She was in denial. Once started upon this path, it was awkward to change her outlook without invoking ridicule from her post-fame, snobbish circle of upper-class acquaintances.

Had she, however, fully embraced her Danish origins; openly acknowledged the funding from her father's fish merchant business; promoted his Jason brand; proclaimed her Hessle Road heritage; and visited Cod Farm to thank the workers for their collection to buy her a car, then she might not have been so rootless and aimless in her famous years. Instead, she might have been much more comfortable and at ease with herself – even admired as a 'local girl makes good'. Indeed, she may have also found an inner peace and purpose during the 1930s.

Throughout much of this publication, I have homed in on some of Amy's personality traits. I began with a list of her characteristics outlined by other biographers such as courageous, foolhardy, impetuous, patriotic and more. I have added other traits such as tomboy, superstitious, escapist, hermitical, unladylike, liar, and meticulous. In my role as a committed historian of Hull's Hessle Road Community, I have probably gone too far in defending the image of the fishing culture and, at the same time, been over-critical of Amy and her motives. For example, denial of her place of birth; distancing herself from Hull; being ashamed of her Yorkshire accent; and even rejecting her parents.

Before drawing this book to a conclusion, I need to redress this imbalance and give Amy credit where credit is due. I need to adopt a more sympathetic approach to her perspective of Hessle Road and society in general. Amy was a woman ahead of her time. 'Ahead' in the sense that she saw through the powerful social expectations placed upon everyone by family, society and culture. Having rejected the charade of humdrum society, she then set about overcoming countless obstacles before discovering what she really wanted to do herself.

Self-determined Amy broke free of various claustrophobic straight-jackets. These arise, sometimes, in the form of everyday clichés like "Don't get above yourself"; "Know your place"; and "Don't rock the boat". She challenged the 'conform from the moment you are born' pressures throughout her life. In Edwardian and post-WWI Britain, social orthodoxy pressed heavily on women. Amy soon spotted that girls, unlike boys, had more restrictions placed on them – such as what they could and could not do. She rebelled against the sexual stereotyping that she first encountered at home and during her school years. For example, she tried to run away from her 'ugly street' almost as soon as she could walk; danced upon the teacher's desk when her back was turned; joined in the lads' rough-and-tumble games that were more in line with her 'tomboy spirit'; showed leadership skills during her 'Panama hat revolt' against her school uniform; cut short her long hair to prevent her mother brushing and curling it to look girly and pretty; and 'twagged off lessons she detested.

Even as an undergraduate at Sheffield University, she did not go along with what the majority of her fellow students did; skipped lectures; and slipped back into her hermit's cave. Office work in both Hull and London drove her crazy. Her low threshold of boredom guaranteed that the treadmill was not for Amy. She saw the rat-race mentality at work and the daily grind to earn a crust. She wanted something higher; nevertheless, she endured stays in London hostels and severe poverty whilst trying to escape from her predicament. She was a survivor of the first-order and adamantly refused to become a victim of circumstances.

Image 37: Amy was always meticulous in supervising the aviation fuel pumped into Jason. Some ground engineers tried to fob her off with cheap, dirty fuel. She knew the consequences if it entered the engine – it could have been catastrophic in the air. At other times, in primitive landing strips, with poor facilities, she filtered every drop herself – even in the baking heat. Courtesy RAF Hendon Museum.

The mould of marriage was never for Amy – even though, at one time, she was desperate to wed. Thank goodness Hans was blind to her talent, inner beauty, future potential, and failed to applaud her fiery nature. She suffered his male ego and arrogance; he exploited her schoolgirl gullibility. He broke her heart, but not her spirit. With the benefit of hindsight, the marriage to Mollison was destined for divorce[132]. Her parents' marriage was not a picture of 'sweetness and light' and her sister Irene's suicide cast doubt upon her 'marital bliss'. Rule-breaker Amy was a role model of non-conformity. Rebel, rebel.

Given that she turned her back on home life, schooling, academia, office work, marriage, and her parents (from time to time), it is not too surprising that she rejected any association with the Hessle Road Fishing Community and her dad's smelly fish dock business. She might well have spotted how exploited were the Cod Farm lasses and other employees of the firm. Added to that, she probably saw the narrow mindedness of community life, petty squabbles, negative attitudes, and the struggle to 'make ends meet'. In effect, Amy was in a state of 'dread-and-fled' about many aspects of life, not just Hull, but the wider social order. Not only did she row her own boat, she was captain of her own one-woman battleship - guns blasting in all directions.

After establishing what she did not want, that left her with lots of time and energy to devote on what she really, really wanted. Eventually, she found her focus in flying. It could have been any area of life - science, religion, politics, stamp collecting or whatever - but aviation became the beneficiary of her love and how to leave her mark upon the world. Not everyone finds what s/he wants in life – Amy did.

Once her mind was set, her willpower was endless. The way she set about gaining her engineering and pilot qualifications was beyond compare. After overcoming those hurdles and with Australia in her sights, there was no stopping her. The speed at which she worked was remarkable and impressive. En route to Port Darwin she applied herself to each and every challenge. One step at a time, she overcame them all. Amy demonstrated that, despite being a troublesome child, a tormented and outcast teenager, and restless adult, if human energy is channelled in the right direction, then victory can be attained. With all her faults and foibles, Amy showed what a determined woman could achieve in the world.

Despite her global fame (or perhaps because of it), Amy found it extremely difficult to obtain employed, paid work as a regular, commercially qualified pilot. It was very much her desire and ambition to be accepted as a professional aviator, but it was still very much a male-dominated closed shop. Equally, some passengers refused to fly knowing there was a woman at the controls. The couple of jobs she did obtain – in 1934 and 1939 - were short-lived.

It was during this period that Amy did some commercial flying: "I was not employed by the company and gave my services voluntarily because I wanted to keep up my flying and get some more hours in my logbook. After three weeks I felt I had had enough [flying London to Paris]. It is one thing having the responsibility of myself only, and quite another to have to worry about six other people"[133].

By 1937 there were hardly any flight records left to be broken. The public, by this time, had become bored with aviation achievements – the novelty value was over. The regular airlines were none too happy with the antics of daredevil pilots having crashes – it was bad for the image of flying. When Amy heard of the death of her close friend Amelia Earhart, lost over the Pacific Ocean on 2 July 1937, she wrote to her mother Ciss "No more flights, so no need to worry. Poor Amelia!"

It is a peculiar twist of fate that both Amelia and Amy died under mysterious circumstances. Both women had a daring spirit and diced with death all their lives. I believe Amy's risk-taking nature was rooted in her Hessle Road childhood where there was a prevailing attitude to "defy the gods and risk the odds" – something Amy did till her dying day. Even at the moment of her death, there was an indirect link with an unknown Hull trawlerman – who desperately tried to save her life.

DATE STARTED	FLIGHT	AIRCRAFT	NAME
5 May 1930	Solo: Croydon, England to Darwin, Australia.	Gipsy Moth DH60	JASON G-AAAH
1 January 1931	Failed Solo: UK to Peking, China. Ended in Poland due to severe winter weather – an ill-advised time to fly east over Russia.	Gipsy Moth DH60G	JASON III G-ABDV ('God Willing')
28 July 1931	Amy + Jack Humphreys: London to Tokyo.	Puss Moth DH80A	JASON II G-AAZV
14 November 1932	Solo: London to Cape Town, South Africa and back.	Puss Moth DH80A	DESERT CLOUD G-ACAB
3 July 1933	Amy + Mollison: South Wales to Connecticut, USA. They crashed but still received ticker-tape welcome in New York.	Dragon I DH84	SEAFARER G-ACCV
20 October 1934	Amy + Mollison: Britain to India (but dropped out of race to Australia).	Comet Racer DH88	BLACK MAGIC G-ACSP
4 May 1936	Solo: Britain to Cape Town, South Africa and return via the East African route.	Percival Gull Six	Name Unknown? G-ADZO

Image 38: These are Amy Johnson's major flights between 1930 and 1936. The table excludes some of her failed attempts (with or without Mollison) due to technical or other reasons. Copyright Alec Gill.

(13) WAR and DEATH: Cut the Engines!

Amy's death is still a mystery. In many respects, she died as she lived – her whole life was enigmatic. Many writers speculate about the shadowy circumstances surrounding her death during the Second World War. Amy patriotically used her flying skills by joining the ATA (Air Transport Auxiliary)[134]. The ATA was not part of the RAF (Royal Air Force), but a civilian operation linked with BOAC (British Overseas Airways Corporation). The Women's Section of the ATA ferried planes around the UK from factories to aerodromes where RAF pilots desperately needed them during the Battle of Britain and later[135]. It was also the case that the war "rescued Amy from the doldrums" after her divorce and there being no more flying records to break[136].

The Woman Engineer journal published a special 'Amy Johnson – In Memoriam' issue in March 1941[137]. The obituary pointed out that Amy accepted "a subordinate position [and] if the country would not use her to the utmost limit of her ability she was determined to be as useful as circumstances allowed"[138]. Reading between the lines, I suspect that, because Amy had a long track record of ruffling the feathers of officialdom, she only received the lowly rank of Second Officer and had to toe the line. It seems she was in the service for less than one year[139].

Wartime did not necessarily mean everyone was glum all the time. A number of retired ATA women looked back at their service as the 'best years of their lives'[140]. Amy had found, at last, a role that made her very happy: she was loyally serving her country; flying planes from one airfield to another; had friendship and purpose within the ATA; and was a popular figure with her fellow pilots and ground staff. She had an inner peace even in the depths of war - with the fall of France, Dunkirk, Battle of Britain, and the Blitz all happening during the dark days of 1940.

Amy spent the last night of her life at the home of her married sister in Blackpool, where Molly's Welsh husband Trevor Jones was the deputy town clerk. This was a precious visit, especially as Amy had just experienced her "lousiest Christmas" – due to bad flying conditions. She was stuck in Prestwick, Scotland - away from her parents' home in Bridlington[141]; but there was a war on and Amy stoically accepted the situation in her usual fatalistic manner.

With the benefit of hindsight of the tragedy to come, these were cherished moments for everyone at No.142 Newton Drive – just north of Blackpool's Stanley Park. Christmas presents were exchanged and the Jones' small daughter "proudly presented Aunt Amy with a large oval mirror bearing a fresco of flowers"[142]. This mirror proved significant in supporting one of many theories as to how Amy came down in the Thames – see later.

Everyone who met Amy during her final week of life commented upon her joyful state. Molly clearly recalled, "she had never seen her looking so happy and well"[143]. Twenty-eight year old Molly would know. She had seen her older sister go through many lows and highs both before and after she became celebrated. In Amy's pre-fame days, there was the bitter family squabbles over Hans; Amy's poor results at university; banned from using the empty family home during the 1925 summer vacation; her failure to become a teacher; her nervous breakdown in Hull; departure to London under a cloud; not attending sister Irene's wedding; Hans getting married; and Irene's unexpected suicide. Then, even after becoming a global star, she was still on a roller-coaster ride with her surprise marriage to Mollison; the massive turmoil when none of her family were invited to the wedding; more world records; anxiety over air crashes in different parts of the world; the inevitable divorce from alcoholic Jim; her various failed love affairs; and the outbreak of war fifteen months prior to her Blackpool visit.

After a happy time at Molly and Trevor's home, Amy returned to the Blackpool RAF Squires Gate airfield on 5 January 1941. It was a depressing Sunday morning and the winter weather was still appallingly bad. A blanket of snow covered the country from the Wash over to Reading and down south to Portsmouth. There were strong north-easterly winds blowing all day long.

Amy was en route to deliver an Airspeed Oxford (nicknamed Ox-Box) twin-engine, four-man trainer plane to Kidlington in Oxfordshire – a 150-mile trip logged as Flight No.V3450. She had mentioned earlier that the plane's compass was faulty. Her official orders from the Duty Officer were NOT TO FLY. The weather forecast of snow showers and bitterly cold winds included freezing fog. One report from Manchester warned of fog and minus 11ºC that morning. Continental Europe was frozen and the anticyclone stretched from Scandinavia south-west to the whole of the British Isles. There was a line of fresh snow showers forecast for the Oxford area[144].

She had, indeed, the ideal opportunity to spend another comfortable night at her sister's home and enjoy a relaxed time with family. Why did she not do that? It seemed totally rational for Amy to opt out of the flight. What is doubly strange is that she even had one of her many superstitious reasons NOT to fly. It was a Sunday. There are strings of general taboos classed as "Never on a Sunday" - never sew, knit, cut fingernails etc[145]. Amy once voiced her feelings about flying on a Sunday[146]. "The holy day" was the only one she believed she had the most accidents[147]. Within the realm of superstition, she had an illogical 'get out clause' - plus an official, logical reason - NOT to fly on such a nasty day.

Image 39: The controversial statue of Amy Johnson stands outside Hull's Prospect Shopping Centre. Local people have mixed feelings and jest that it looks like a Jelly Baby, the Michelin Man or something bought from a garden centre. I was pleased to see that the plinth listed her record-breaking flights and her role in the ATA. Finch (1989) highlighted that there was "so little to preserve the name of one of the city's great heroes" and it took an outsider (Irishman Leo Sheridan) to initiate the memorial (p73). The final, bold statement on the plinth declares, "MAY HER FAME LIVE ON". Copyright Alec Gill.

There is, however, an unseen hierarchical order within the world of superstition. A more powerful taboo sometimes over-rides a lesser one. Near the very top of the pyramid is the folklore belief "Never Turn Back". It could be that this Hessle Road seafaring superstition kicked-in and prevented Amy from taking the easy option. Once started out on a journey, it was very bad luck to return home. For example, it was an ominous sign if a trawler suddenly returned to port, for whatever reason.

One Hull example of this belief in action was associated with the sinking of the Endon (H.161) in January 1933. This trawler had only been at sea for one week when she unexpectedly returned. The ship's mate was washed overboard and the vessel came back. Some crewmen were disturbed and refused to sail again. Nevertheless, the Endon did set off back to the North Sea fishing grounds just before Christmas 1932. A week or so later, the St.Kilda (H.355) spotted the Endon aimlessly adrift with all her trawl gear down and lights ablaze in daylight. Three St.Kilda men rowed over to the helpless vessel only to find that there was not one soul on board and the lifeboat was missing. It was spooky - like the Marie Celeste. Superstitious rumours soon made the connection between the fact that the trawler had turned back causing bad luck for the cursed ship and its doomed ten crewmembers.

Because it was bad luck to turn back, Amy took off – even against all sensible advice. This very superstition had perhaps kept Amy going forward during her flight to Australia. No matter what struggles were encountered, she never turned back and luckily made it to her final destination. Who knows what goes through anyone's mind when in an emotional life-and-death situation? Superstitious individuals rarely understand their own irrational behaviour – nor do much to change it, even in the face of harsh reality.

Yet another impulse – more patriotic and rational – may also have been at work within Amy's mind. She was the sort of person who placed duty first. She made this clear in her article 'A Day's Work in the ATA' – posted to the editor of The Woman Engineer journal on 1st January 1941 (just four days before her death). She expressed the desperate need to keep aircraft production moving – despite any bad weather. Prophetically, she wrote: "In the winter months, aircraft production and repair work go on unabated, increasing every day, but the bad weather holds up delivery of machines from factories to aerodromes, and numbers of planes accumulate, making no mean problem for the organisation staff, as, somehow or other, in spite of the weather, the machines must be dispersed in order not to present too alluring a target for the enemy"[148]. This was efficient Amy, who was always highly motivated to 'get the job done'.

Pauline Gower, Head of the ATA, wrote of Amy that, "whatever the circumstances, however she was feeling, the job was done; and the conscientious manner in which she carried out her duties was an inspiration to all those who worked with her"[149]. This strong sense of 'duty first' was later confirmed by Molly who had "pleaded with her sister to stay, but Amy had probably already notified Squires Gate aerodrome of her intended time of departure and seemed possessed by her own concept of duty. She replied, "It's an RAF plane and the boys are waiting for it. I've got to go. Don't worry, I'll fly over the top of the clouds and smell my way to Kidlington"[150].

Image 40: Whenever Amy put her mind to a job, she did it with meticulous attention and care. She showed this ability during her flight to Australia; but especially applied it during her ATA duties. The job came first. The delivery of planes was paramount. The supply chain must never be broken – especially in the depths of winter. Whenever Amy touched down at any RAF station, she was always very popular with the pilots and ground crew who gathered around to see her. Equally, she was happy to chat with them about flying and aircraft maintenance. Courtesy Hull Daily Mail.

She flew off at 11:49 hours with a full tank. As over the Timor Sea, she had her own strategy for dealing with ugly-looking clouds – go around or above them. Yet within the ATA, there was the expression "Killer Clouds" – most of the 136 casualties were due to bad weather conditions[151].

Like the Endon, there are many mysteries around the death of Pilot Johnson. Whilst researching Amy, a variety of people, uninvited, put forward their own forceful views about how and why she died. The most prominent opinion was that the British deliberately shot her down because she was carrying a German spy over to occupied France. Not only that, this spy was her lover![152] Another version on this theme is that the British shot her down accidentally because she did not radio back with the correct secret code for that day.

These curious stories are unsupported by any tangible evidence. As some journalists say, "why let the facts get in the way of a good story". I do not intend to spend masses of time analysing a variety of speculations. Instead, I am happy to draw upon the research of biographers such as Smith (1967), Luff (2002) and Gillies (2003). The following account of Amy's death is largely based upon their material.

The main 'shooting down' story arose from 83-year old Tom Mitchell in February 1999. He was a gunner with the Royal Artillery (207 Battery of the 58th Heavy AA Regiment) on the day when Amy's plane came down. Gillies (2003) checked all aspects of his account and interviewed Tom himself. She pointed out (a) Amy's Oxford aircraft had no radio on board (few did when they left the factories) – thus, no conversation could have taken place; and (b) Tom's gun battery was "stationed beyond the Isle of Sheppey, twenty miles away from the point at which Amy's plane went down – a distance well outside the guns' range" (p342-43). Tom had carried a false guilt for 58 years because a senior officer told him to keep quiet under the Official Secrets Act. Furthermore, David Luff established that Tom's battery at Iwade, in Kent, fired its guns between 19:00 and 23:30 hours on 5 January 1941 – three-and-one-half hours after Amy's plane came down in the Thames[153].

Anyway, back to Amy over a densely, cloud-covered Britain on that fateful Sunday afternoon. It was nearly four hours since she left Blackpool. Under normal, good weather conditions, the flight should have taken around 90-minutes. She was desperately running out of the plane's original 32 gallons of fuel and around 140 miles off course in an easterly direction. With no choice, she came down through the very thick clouds to establish where she was. She then saw Convoy East 21 in the Thames Estuary. It comprised seventeen merchant ships escorted by eighteen naval vessels. Amy bailed out, hoping that one of the British ships would pick her up. It was 15:30 hours in the afternoon as the winter light was beginning to fade.

It is hard to believe, but Amy had never made a parachute jump before in her life – it was her first and final one. As per standard procedure, she locked or 'trimmed' the plane into a clockwise, spiralling motion as it circled downwards[154]. Before jumping, Amy pushed out her two leather bags of belongings. The air temperature was below freezing. The blast immediately cut through her face and body like a cold steel knife. She slowly drifted down through the fog and sleet into a watery abyss.

As she plunged into the icy water, her traumatised body soon began to numb. The complete shock to her metabolism was horrendous. Her lungs almost exploded. Beneath the waves, bubbles gushed from her mouth. She repressed the instinctive urge to breathe in. Everything was in slow motion. Amy gasped for air when she broke the surface. She swallowed seawater and instantly spat it out. Every breath was sharp and painful. Snow continued to fall. Blood rushed to protect her brain from becoming unconscious. Numbness quickly advanced to every finger, every toe. The pain was beyond pain – it was insane. Her hands turned white and stiff. Her greatest immediate risk was not drowning, but freezing to death. She was soaked to the skin, but that was nothing compared to the agony of the bitter cold.

The swell and current swept Amy along, helplessly, in the powerful Thames tide. A large ship loomed through the gloom of the late dusky afternoon. She could see its lights – there was hope. She could hear its spinning propeller carving through the water – there was death.

Fate decreed that she landed near the escort vessel HMS Haslemere – a former English Channel cargo ship built in 1925 and classed as a Barrage Balloon Vessel[155]. The ship had a large balloon flying overhead as part of the protection for the convoy. As soon as a parachutist was spotted about 800 yards away, the ship headed straight in that direction to effect a rescue. The state of the weather at the time was sleeting; it was bitterly cold and the sea had a heavy swell.

Whilst steaming toward the unknown pilot, the Haslemere ran aground on a sandbank. When a vessel runs aground, it is not a quick or easy operation to work free – skilled seamanship is essential to avoid making a bad situation worse. Lieutenant Commander Walter Edmund Fletcher RNR was fully qualified and experienced to deal with such a problem. He "ordered a starboard alteration of course to manoeuvre ...leeward. Almost immediately, the quartermaster reported that the ship was refusing to answer her wheel. There had been no perceptible shock but she had obviously run up on the sand. Now, with the keel beginning to bump, Fletcher rang slow astern on both engines and began to claw his way, inch by inch, out of danger"[156]. The task took all of the attention of the officers on the bridge - they had to focus upon every moment and movement to ensure the safety of their vessel.

The manoeuvre took time before the ship could continue its initial rescue attempt of the parachutist. In the confusion of war, reports differ widely as to the movement and direction of the ship's propeller. There were, obviously, three possibilities: reverse, forward or static. Witnesses vary in their accounts. The truth might never be known. I believe the Haslemere was still endeavouring to get off the sandbank and the propeller was spinning in reverse – sucking the helpless parachutist toward the tumult.

Ordinary seaman Nicholas Roberts heard a woman shouting in the water and twice managed to throw a heaving line to her. She was almost twenty yards to the stern of the ship. Frozen stiff, Amy was unable to reach the lifelines. Next, seaman Raymond Dean bravely mounted the rubbing board near the stern of the 756-ton Haslemere. Holding on with one hand, he tried desperately to reach her with the other. The ship was lurching up and down. The nearest Amy came to him was five feet – just beyond reach and rescue. They must have looked into each other's eyes.

Over and over again, Amy pleaded "Hurry, please, hurry". Luff stated that the stationary ship was heaving with the strong tide causing the stern to rise and fall in the heavy swell[157]. The generally accepted wartime (official) accounts suggested that the propeller was not spinning at that exact moment.

In 1991, some previously unknown information came my way. The needless tragedy was avoidable. I certainly do not have the full picture of what took place during the chaos of war – no one has. However, as a historian of the port's fishing heritage, I was given a snippet of information relating to Amy's death. This came from the son of Harry Gould (Senior) who was aboard the ship involved in a failed attempt to save Amy - HMS Haslemere[158].

The evidence of Harry Gould is hearsay, incomplete and clashes with some of the published events surrounding the tragic incident – equally, it fits in partially with other, more recent, related evidence. Nevertheless, as an interested party, I feel I must put it forward for future historians to take into account. Even if it takes one hundred years to solve, every bit of evidence has to be weighed. It is within this context that I submit the Gould Family material.

Hull trawlerman and Hessle Roader Harry Gould was also at the stern standing by to help fellow seamen Roberts and Dean in their rescue attempt. As he watched, he soon realised that the engines were in reverse and the helpless parachutist was drifting perilously close to the ship's slashing blades. Harry could hear the calls for help and knew that those on the bridge had no idea what was happening out of sight at the stern. In a desperate bid to help, he frantically dashed toward the bridge yelling and repeating, "Cut the engines! Cut the engines!"

Unfortunately, an officer on the bridge ignored the shouts of a mere trawlerman from the lower ranks. He continued running the engines astern, and called back, "Don't you tell me what to do!" In the following vital seconds, poor Amy was dragged under the stern of the vessel and her body cut to ribbons. Seaman Roberts later stated, "The ship was heaving in the swell and the stern came up and dropped on top of the woman. She did not come into view again"[159].

More recently, fresh evidence, from another eyewitness, has come to light. I am keen to cite this material because it corroborates that someone aboard the Haslemere did dash toward the bridge shouting something out – even if the context was slightly different. A television investigation partly confirmed the Harry Gould version of the tragedy.

In October 2002, the BBC TV Inside Out team claimed to have traced a new key witness who spoke out for the first time after sixty years of silence. Derek Roberts (not related to the Nicholas Roberts mentioned above) was a clerk in the Thames RAF Flight Office on 5 January 1941. His friend RAF Corporeal Bill Hall was aboard HMS Haslemere and saw what took place. Bill was still shaking from the incident when he reported what had happened. Derek typed up the account as his friend spoke; Bill approved and signed it; it was then sent up to their Flight Commander. Nothing was heard of his testimony again and Bill was not called to testify at any official inquiry. Derek has long suspected "an official cover-up over Amy's death"[160].

The significant aspect of Bill Hall's report was that he stated, "They threw her [Amy Johnson] a rope, but she couldn't get hold of it. Then someone dashed up to the bridge and reversed the ships' engines as a result of which she was drawn into the propeller and chopped to pieces".

Image 41: Harry Gould Senior is here with his wife Violet at their home No.96 Regent Street, Hessle Road in the 1950s. He was "mentioned in Dispatches" in May 1941, but no specific details were given in the London Gazette LT/JX 219467. I am not sure if the child squeezed in between the couple is Harry Gould Junior – my informant of this vital story. Courtesy Harry Gould Jnr.

I am obviously keen to relate this evidence. For the first time, it substantiates that someone did dash up toward the bridge; but I firmly believe that Harry Gould repeatedly shouted, "Cut the engines" – not reverse them. The engines were already spinning in reverse. The Haslemere was still trying to manoeuvre off the sandbank when Amy was chopped to pieces by the spinning blades.

Midge Gillies also strongly argued that the ship's propeller mangled Amy's body and cut her to pieces[161]. Had the stern of the vessel or its static propeller simply come down on top of her and knocked her out, then the chances of Amy's whole body being found would have been much higher. Her body was never found, nor fragments of her uniform. I agree with Gillies' opinion: Amy's body was chopped to pieces because the engines had not been stopped. The propeller would have stopped if someone in the wheelhouse had heeded Harry Gould's plea.

For some reason, the Gould evidence was not available at the official inquiry in December 1943, almost two years after the incident, but it certainly deserves some consideration. Harry and his son had no reason to lie and their details more or less fit with other accounts of Amy's death. Furthermore, it explains why her body was never found.

The Gould account, however, did not identify the superior officers on the bridge – and I am certainly not making any allegations. The David Luff (2002) biography explored in greater detail what happened after Amy had been chopped to pieces and the strange behaviour of the ship's captain. Lieutenant Commander Fletcher bravely dived overboard to make a desperate and futile rescue attempt. The media at the time, based upon only four eyewitness statements, speculated that a second parachutist also jumped from Amy's aircraft – Mr.X.

Haslemere's captain dived into the freezing water in the hope of saving the unidentified 'second person'. Later evidence suggested that Amy's travel bag was mistaken for a person's head bobbing up and down – kept afloat by an air pocket. Tragically, he suffered hypothermia in the attempt. He never regained consciousness after being dragged out of the Thames and died. Within four months, May 1941, he was awarded posthumously the "Albert Medal for gallantry in trying to save life at sea"[162].

Luff, however, injected an interesting critical note to the Fletcher dimension. He quoted the Admiralty's Director of Navigation (unnamed) who stated that, "it is wrong in principle – especially in war – for any Commanding Officer to abandon, even temporarily the command of his ship, by himself diving overboard for the purpose of saving life at sea"[163].

Why did the captain jump into the water and abandon his command? How does the Gould version fit into the above drastic incident? Could it be that perhaps Captain Fletcher regretted that someone on his bridge did not heed Harry's pleas to 'cut the engines' and so, on an impulse, he made a desperate rescue attempt to save the life of the presumed second parachutist? We will never know. I certainly hope that more historical evidence will still come to light.

Image 42: This is one of several Bethel Boards "In Loving Memory of Loved Ones Died at Sea". These were once housed in Hessle Road's Fishermen's Bethel. This list is far from complete – only a fraction of the 6,000+ fishermen lost from the port. Nevertheless, for those fishing families whose relatives are named, it is the only tangible record of their lost loved one – the nearest thing to a tombstone. These boards are now on permanent display in the shore-based museum next to Hull's heritage trawler Arctic Corsair (H.320) – moored in the River Hull off the port's Old Town. Copyright Alec Gill.

The news of Amy's death devastated the Johnson Family. They had now lost a second daughter prematurely – this time to the sea. In a cruel and perverted twist of fate, Amy's plunge and plight paralleled that of countless trawling families who had also lost their offspring to the sea. Will and Ciss now shared the same trauma of thousands of Hull fishing families (Image 42). That is, there is an affinity, an emotional empathy, for anyone who losses a loved one at sea when no body is ever recovered. There is no funeral to attend; no grave into which the body is lowered; no tombstone where flowers can be laid at times of a special anniversary; yes, there might be a memorial service, but no mountain of memories can replace a physical place. The letters R.I.P. stand for Rest in Peace; but there is no peace of mind for relatives and friends left behind unable to lay to rest their departed. This bond binds together people whose loved ones have only a watery grave beneath the waves.

There is a powerful Hessle Road superstition: "Never wear pearls. If you take something precious from the sea, then the sea will take something precious from you". For decades, the Johnson Family had certainly made their fortune from the fish of the sea and prospered from generation to generation. Thus, it came to pass for the Johnsons that the sea now claimed the life of their precious Amy in return for all the souls of the fish taken for the benefit of the AJK empire. The sea now claimed an extra special 'pearl' from the Johnson Family. Perhaps deep down both Will and Ciss realised that "the price of fish is paid in human lives".

Amy's death was obviously a very gory ending to such a colourful life. There is a chance that certain evidence about Amy's death was officially supressed in a time of war so as not to dishearten those involved, especially her family and the British nation as a whole who saw Amy as their wonderful hero? The Johnsons lived the rest of their lives in the shadow of their missing aviator daughter[164] (Image 43).

The news of Amy's death was all over the national newspapers. The Trustees' Minute Book (1910 – 1951) for the St.George's (Road) Wesleyan Methodist Church recorded the following entry on 30 January 1941 at 8 pm "It was agreed that it should be placed on record: A letter of sympathy had been sent to Mr & Mrs J W Johnson the parents of a late scholar of our Sunday School (Miss Amy Johnson)"[165].

This very brief note touched upon a deep sense of loss. A girl who was born and bred, played and sang in the streets of Hull's Hessle Road had just been killed in war serving her country. Amy had danced upon the world stage, she was the darling of the nation she loved, and now she was struck down.

Ironically, Amy predicted how she would die. She was chatting to a fellow ATA pilot as both pondered the death of a colleague in a plane crash. Gloomily and prophetically, Amy declared, "I know where I shall finish up – in the drink…A few headlines in the newspapers and then they forget you"[166]. Sadly, she was right about being lost at sea, but fortunately very wrong about being forgotten by the nation or her homeport.

Image 43: Amy's father Will Johnson - possibly at his post-WWII home near Beverley - proudly shows off his collection of memorabilia in 'Amy's Den'. Will was a tidy man who was keen to ensure that Amy's artefacts were carefully preserved. For a long time I wondered why nothing much remained in Hull. An Alderman of Hull City Council, who knew the Johnson Family well, recently contacted me. He told how Will had initially offered Amy's material to the Curator of Hull Museums. The bureaucratic official only agreed to accept them on condition that the family paid all costs, transport and fully insured the items. Upset and angry by such a negative, narrow-minded response, Will offered everything to Sewerby Hall, Bridlington who gladly accepted the generous gifts without any pre-conditions. Courtesy Hull Daily Mail.

(14) FUTURE: May Her Fame Live On

For many decades, Amy seems to have had little prominence in the celebration of aviation heritage - nationally or locally. Occasionally, her name is recalled during certain anniversaries and there was a second-class Royal Mail stamp in her honour during 2003 (Image 44) - but nothing of any real significance until recently. Suddenly, there is much renewed interest in the Amy Johnson story with a wide variety of colourful projects planned for the years ahead.

Captain Amanda J. Harrison, based near Oxford, has (or had) plans to retrace Amy's solo flight from the UK to Australia, using a Tiger [not Gipsy] Moth biplane. Her scheduled take off was set for the 19 May 2015 and expected to take 16 days, but this date seems to have lapsed. I wish her well when (or if) she does embark on her 'Australian Adventure'[167].

Secondly, Tracey Curtis-Taylor did actually take off (1st October 2015) in her 1942 Boeing Stearman open cockpit bi-plane called the Spirit of Artemis. She wisely decided to make it a three-month trip with 50 re-fuelling stops, and covering 13,000 miles taking in 23 countries. She avoided war zones and took in famous sites like the Taj Mahal (India) and Uluru (Ayers Rock, Australia). Tracey's was a leisurely route and timescale, but her venture still held potential dangers from unpredictable weather and engine failure. This is another brave venture worthy of praise and support. Tracey landed in Sydney in early January 2016 and marked the 75th anniversary of Amy's death[168].

Thirdly, Rick Welton is organising 'a festival of the arts and engineering inspired by the life and achievements of Hull's aviation heroine, Amy Johnson' - between July and October 2016[169]. This involves a wide range of Hull-centred events and activities. Four broad themes are: (1) the life and flying achievements of Amy Johnson – telling her story; (2) Women's equality and liberation – Amy as a role model to improve female opportunities in the worlds of engineering and science; (3) Freedom of the Skies – the role of aviation to improve international communications, making the world a smaller place for more people; and (4) The Culture of the 1930s – music, literature, fashions, visual arts, domestic life, and architecture.

Image 44: The Royal Mail issued an Amy Johnson commemorative stamp in May 2003 - the centenary year of her birth in St.George's Road. Gradually, Amy is receiving the local, national and global recognition she fully deserves. Courtesy Royal Mail.

Rick's other plans are for cultural exchanges with the landing places where Amy re-fuelled en route to Australia such as Istanbul, Baghdad, Calcutta, Rangoon, Surabaya, and places in Australia. Within Hull itself, there will be a series of high-quality exhibitions, concerts and productions during 2016. At street level, sculptor Saffron Waghorn will create and display about 100 decorated model moths (insects) around the city and beyond. Local schools will also be involved in many activities.

Fourthly, there is the Amy Johnson and Herne Bay Project that hopes to find the wreckage of the Airspeed Oxford aircraft about 12 miles out into the North Sea in the sand banks off the Kent coast in the Thames Estuary – led by Jane Priston[170]. This work is gathering lots of local support throughout Kent. The search is in conjunction with the Canterbury Divers who will conduct a detailed survey of the seabed where the aircraft might be located. It is looking for 'a needle in a haystack' but the team's motto draws upon Amy's own words "believe nothing to be impossible". Good luck to them because every shred of evidence will help clear up one less mystery surrounding Amy's death.

Talking about her wreckage, there is the growing view that Amy did not bail out, but actually performed a controlled landing of her plane on to the surface of the Thames. In 1989, Paul Crawley proposed that Amy landed her plane in the estuary[171]. This is certainly a perspective worthy of further research.

Aviator Tracey Curtis-Taylor also subscribed to the view that Amy did not parachute from her Airspeed Oxford. Instead, she controlled the aircraft onto the surface of the Thames, lifted out her two bags, and entered the water as the wreckage began to sink. Tracey's reasoning is because Amy's mirror gift she received in Blackpool was unbroken. Had the mirror dropped from the aircraft before Amy did a parachute jump, then it would have shattered on impact. The article concluded, "Ms Curtis-Taylor believes Amy was killed when the boat attempting to rescue her sucked her into its propeller"[172].

A summary of Amy's childhood on Hessle Road and her death in the Thames Estuary could be that she was a girl born on the banks of Mother Humber and died in the arms of Father Thames. Amy Johnson - the person, the legend - will certainly live on for many decades to come. She is a wonderful role model for many to follow, especially young women. If Amy was trying to 'escape' from her family's past, then she was bound to fail – no one can flee the inner self. It seems she never learned to love herself for who she was. Who knows, had she done so, she might have lived into old age and really advanced the cause of women engineers.

I must end this work with Amy's own words. Just four days before she died, Amy typed a cover letter to Caroline Haslett the editor of The Woman Engineer journal. It was dated 1st January 1941 – the start of a brand new year – it was a letter of hope for the future, written straight from the heart to a close friend. For me, it reflects not only her sense of humour, but also the fact that she was a born-and-bred Hull girl who had really absorbed the local fishing folklore beliefs of having a lucky outlook. Deep down within her heart she was still that playful Hessle Road tomboy. Her final sentence and wishes in the letter were:

"I hope the gods will watch over you this year, and I wish you all the best of luck (the only useful thing not yet taxed!)"[173].

Image 45: The Amy Johnson Cup for Courage, displayed at Hull's Guildhall, is a marvellous gift that Amy left to her hometown. In 1930, Sydney schoolchildren gave Amy a purse of gold sovereigns worth £25 (valued at £1417 in 2015). She used the gold to create this cup to encourage Hull children to be brave and daring in their young lives. It is awarded to a young person from within the city's boundaries (under 17 years of age) for an act of bravery – such as saving a life. In stark contrast to the few gold nuggets her father brought back from his youthful trip to the Klondyke Gold fields, his daughter hit the jackpot with her own daring and courageous odyssey to Australia. Courtesy Hull City Council.

(15) APPENDICES: Introduction

Amy Johnson was not only an aviator, but also an author. The problem is, however, that much of her writing is hidden away in obscure places: journals; a single autobiographical chapter in a rarely seen publication; newspaper articles; and, as a solo author in 1939 (Sky Roads of the World), her book received mixed reviews and it was overshadowed by the Second World War.

In order to find Amy's source material, I had to conduct an extensive search in out-of-the-way places such as the Reference Section of the Hull History Centre (ticket-only access); Carnegie Heritage Centre; and the RAF Museum in Hendon. On one occasion, it was a matter of good luck. Peter Nicholson, the owner of Marsh Nicholson Furniture Shop on Hessle Road, happened to mention that his dad had been at the Boulevard School when Amy was there too: Did I want to borrow a copy of The Boulevardian in which she had written an article? Yes, of course, I did – thank you! It was an amazing little piece called New Year's Day in China (Appendix 15C).

Given the unseen side of Amy the writer, I decided to bring a selection of her excellent writing into the public domain. Thus, readers will be able to appreciate Amy as an articulate author. Her writing highlights her personality, sense of fun and adventurous spirit. Furthermore, readers will also openly see how and from where I have cited Amy's words - without anyone having to track down the source material from stock shelves in reference libraries.

When I began to read Amy Johnson directly, I was impressed by her openness and honesty - her "rose-coloured glasses look at life". She has a friendly style and her personality shines through her words. Another appealing aspect is her wit and humour.

I was tempted to insert a range of EndNotes at many stages whilst reproducing Amy's writing - to link directly what she was saying to my text in the main body of this book. Immediately, however, I stopped and questioned that approach: Why should I inflict my dictatorial views on the reader? Why should I attempt to impose my narrow interpretation on her words? Moreover, it would be disruptive to each reader's pleasure of being absorbed in her language and feelings.

Instead, I decided to leave her words largely uninterrupted so that each reader can enjoy Amy's writing at first hand and come to her/his own conclusions. After each piece, however, do I add some reflective comments - sometimes linking them to the main body of my book.

There is another reason I am eager to set before you Amy's text. What I found with some biographers is that they based certain chapters directly upon Amy's own writing, but omitted to cite their original material. The impression was given that they perhaps had access to some private diaries / papers and were reluctant to divulge their origins – as if it was confidential. Some biographers lifted whole paragraphs verbatim from Amy's autobiographical chapter Myself When Young – without ever citing her. This annoyed me and I thought it was unfair not to give Amy the credit.

I have imposed some slight factual details within Amy's text. Where I have done this, I have inserted my comments within square brackets. When she mentioned borrowing £50 "from a friend", for example, I could not resist inserting [Hans Arregger – her Swiss lover]. Therefore, in that respect, I have transgressed my self-imposed rule of not influencing the reader. I suppose the historian in me felt impelled to 'set the record straight' and not allow Amy to smooth over certain cracks!

Occasionally, I have also used square brackets to show where I have inserted the odd word to help the syntax of a sentence. Amy's punctuation of her text has been left as she did it. Fashions change in punctuation style, but I decided to leave some of her long sentences untouched.

Please be my guest and enjoy – at first hand - Amy the author, beginning with Appendix 15A: Myself When Young. Appendices 15A, 15B and 15C are reproduced in their entirety; whereas the other pieces (Sky Roads of the World and The Woman Engineer) are extracts. Hopefully, these Appendices will whet your appetite and encourage you to read Amy's words directly.

Being a photographer, I have inserted some images alongside Amy's writing. So please bear in mind that the illustrations and captions have been 'planted' by me. I hope they enhance her words and add to the enjoyment of your reading.

##################

(15A) Myself When Young
by Amy Johnson

Appendix 15A: JOHNSON, Amy. 1938. Chapter. Amy Johnson. Myself When Young - by Famous Women of Today. Edited. The Countess of Oxford and Asquith. London. Frederick Muller Ltd. 131-156.
[NOTE: A photocopy of this chapter is available at the Hull History Centre, Reference L920 JOH – 54/13/6]

"I'se a Queen, I'se a Queen," boasted a tiny mite of three years old as she strutted sturdily up and down the nursery in her comfortable Yorkshire home. With childish mind brimful of ambition and adventure, she stealthily snatched her bonnet and, when Mother's back was turned, slipped out of the open front door and unsteadily trotted down the ugly suburban street. Choosing a house which attracted her with its new coat of bright green paint and front garden of gay spring flowers, she pushed open the gate and beat on the door with small dimpled fists.

"I'se run away to be a Queen," she confided to the woman who opened the door. "Running away indeed! You'll go straight home to your mother," was the unsympathetic response, and without more ado this spirit of adventure was unceremoniously carried home screaming and kicking and vowing vengeance.

Some twenty-five years later, a young woman glanced down with quiet amusement at the writing on a postcard which had been placed on the dressing-table in her cabin. She was cruising along the coast of Africa and the ship had just put in at Port Elizabeth, famous for its Snake Park. The card was a picture of an immense negro with a mighty cobra entwined round his athletic body. Both hands held a wicked-looking head, poisonous fangs unsheathed a mere few inches from his face. Underneath was written this inscription: "From Johannes, King of the Snakes to Amy, Queen of the Skies." Smiling to herself she remembered her childish boast of so many years ago. "Well, I always said I would be a Queen!"

* * * * * *

Many years later still, I sit at my desk and think back into the past. I drag out memories one by one into the light of day, gently dust and lovingly put them away again, for with not one of them would I part.

So much has happened during these years that memories crowd thick and fast, the more colourful trying to blot out those of more faded hue, so that I have to look carefully into the shadows to drag out perhaps some trifling incident, some half-forgotten word or dream which has nevertheless served its turn in giving shape and substance to my life.

Vivid pictures the years have failed to dim are [of] my mother in [her] bright red and white checked blouse, sleeves rolled to the elbow, slender fingers and small brown arms buried in foamy soapsuds as she scrubbed minute frocks and petticoats at the open window of the tiny scullery. Singing cheerfully as she works, every now and then she calls to me as I play in the garden, getting as dirty as only I know how and heedless of the work and trouble I am giving her. For it was not at the age of five years that I realized how young and pretty my mother was, and that, married at eighteen, she was spending the best years of her life looking after troublesome me and my sister Irene, fifteen months younger than myself.

My newly washed underwear hanging on the line, I would wheedle Mother into the drawing-room and get her to tell me a fairy-story and then play the piano so that I could dream it. This was my favourite amusement for years. Mother played like an angel and I would lie on the sofa with my eyes shut living a wonderful secret life of my own, full of exciting adventures in which I was always the heroine and the end was always happy and satisfying.

I adored fairy-tales and from the moment I could read I would lie sprawling on the hearthrug, stuffing my head with stories of knights in shining armour, beautiful princesses with red hair who fell in love with handsome young peasants they found trespassing in the Palace woods and who always turned out to be Princes in disguise. The thousands of fairy-tales I read have probably had a great influence on the formation of my character, for even to this day I tend to look at life through rose-coloured glasses, always see the light before the shade, the good before the bad, and expect a happy ending. Disillusionment has been hard and bitter, but has never been complete, for the habit formed in childhood always in the end transforms the ugly toad into a handsome young Prince.

A particularly vivid memory, and one of my very earliest ones, is of my mother patiently explaining to me, almost a baby then, that it was simply not done to help oneself to other people's belongings, no matter how much I might want them. The well-deserved lecture arose from my attempted theft of an apparently irresistible black china cat sheltering under a red umbrella, which reposed on the mantelpiece of my granny's house. It seems I made three distinct efforts to steal it, which must have decided my much distressed mother that my future was to be a career of crime. I am afraid that I still often want things I see on other people's mantelpieces, but I have learned that there are other, and legitimate, ways of getting them.

My clearest memories of my father in my earliest childhood days are of him romping on the floor, as a bear, encouraging me to throw away my dolls and play with trains instead; of visits to the sweet-shop at the top of the road where a pennyworth of peppermints would console me for my grief that I had forgotten, once again, to put out his slippers to warm in front of the fire; of the penny he would give me when I pulled out a hundred daisy roots from the lawn, or repeat from memory a hymn I had learned in chapel on Sunday. To my mind, my father knew perfectly how to deal with small children in chapel. The longer the sermon the better I used to like it, as he kept his pockets full of lovely luscious chocolates, in pretty silver and mauve papers; and then there was always the penny to be earned whenever I could be bothered to learn some verses.

Of my little sister Irene, who most tragically died at the age of twenty-three, I have the softest and tenderest memories. Pretty and always very delicate, it seemed my job to protect her. For hours I would sit at night in the dark, with the firelight leaping and making terrifying faces on the wall, too frightened almost to move, but holding on to my sister's hand protesting with shaking voice that there was nothing at all to be frightened of.

The games we played together were always unusual and of our own invention. We both shared a hatred of the ordinary, of the conventional. It was always quite enough for us to be told we must not do something, to want to try it, and to be told it was impossible, to want desperately to succeed.

One of our favourite games was to make a tour of the nursery, or the drawing-room, without once touching the floor. You can well imagine the damage to cushions and furniture as we leaped from the sofa to the top of the piano, thence to the mantelshelf, and on one historic occasion a trapeze act on the gas chandelier sent us both to bed for the rest of the day. Who knows that perhaps from this early game I caught my passion for hopping around in the air without touching ground more often than necessary!

Inevitably our carefree life had to come to an end some time, and school days started. I was sent to a series of private schools whilst my sister stayed at home waiting longingly for me to come back to play.

Memories of my early school days are vague and confused. Apparently in those days I had a greedy passion for acquiring knowledge, which the limited education available then entirely failed to satisfy. Run by nice old ladies, whose only qualifications to teach the young seem to have been high-class birth and breeding, these schools no more liked me than I liked them [Appendix 15B].

Headstrong, and probably spoilt, with an insatiable desire to know everything, I seemed to have pushed my way through each school in turn, enlivening the hours spent amongst girls twice and three times my age by pranks of incredible foolishness and foolhardiness. It was, perhaps, my tomboy spirit which, thank God, saved me from the danger of becoming a blue-stocking. I brought homework home with me for the sheer love of doing it, and gave my teacher hours and hours of extra work correcting interminable sums of algebra and arithmetic. I learned a queer mixture of elementary and advanced subjects, and when I finally left private schools to enter a secondary school, at the age of eleven, I could boast of a smattering of physiology, algebra, geometry, trigonometry and biology. This knowledge, coupled with the fact that for years I had been lording it over girls so much older than myself and had played elder sister so successfully to Irene, gave me a good superiority complex and headed me straight for the fall which was coming to me.

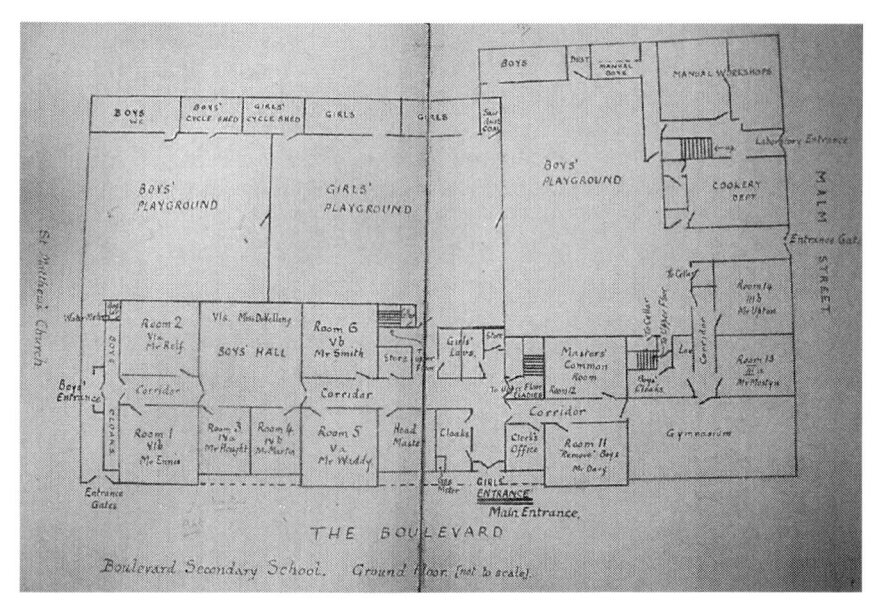

Image 46: Ground floor plan of the Boulevard Municipal Secondary School - probably as it would have been when Amy first attending c1914. Courtesy Old Kingstonians' Association.

In my eleventh year, my mother wisely decided that a secondary school would do me a world of good, so she made application for me to enter the Boulevard Secondary School in Hull. A smart maroon tunic was made for me, a straw "banger" purchased and put on my unwilling head, a too, too clean white blouse made me feel most conspicuous and self-conscious, and, to add to my sense of strangeness in this queer uniform, a white drill shoe-bag, with initials neatly embroidered by myself in maroon, was put in my hand, inside which reposed a brand new pair of gym shoes. Resplendent in this get-up, one fine autumn morning when I had just acquired the ripe age of twelve years, I set out, holding on to my mother's hand, for my new school.

It was the first day of term and activity was intense. Waiting our turn, along with many other new girls, I took stock of my surroundings. We were sitting in a large bare hall, sparsely furnished with some desks, a table and a high reading-desk, from which the lesson was read by the Prefect, at morning prayers. I was soon to know intimately every corner and stick of this gloomy cheerless place. Even this first morning I had more than enough of it before our turn came to go into the Headmistress's room. We were the last, and almost never got there at all as, apparently, we were unexpected!

Our entrance form had been mislaid or lost, or maybe had never been posted but, whatever the reason, my name was not on the list, and it looked as though I should have to go home again. My mother, rightly indignant that all her work and expense in acquiring my outfit should go wasted, insisted that I should be taken. Eventually she won the day, but it was found that I should have to go into a class with girls younger than myself as there was no room in the class of my own age and standard.

From that moment, my school days became grand fun. I found that I knew much more than the rest and that I hardly had to do any work to be easily at the top of the class. I therefore developed excessively lazy habits, and, as you can probably guess, my talent for mischief found full play during many leisure hours. It would weary you to tell of these school days in detail but, as they occupied seven of the most valuable years of my life − I was at school until I was nineteen − I will relate some of the highlights, and those incidents which seem to have had most influence on my future career.

First and foremost, the school being co-educational, I rapidly developed a great interest in the opposite sex. I took a fierce pride in my popularity with my boy class-mates and became as big a tomboy as any of them. I played all the boys' games, boasted of being the only girl in the school to bowl overarm at cricket, played hockey with an intense zest, and scorned tennis. Gymnastics were my strongest point, and besides the classes at school, I went twice or even three times a week to the gymnasium of the Young People's Institute, which had one of the best equipped gymnasiums I have seen anywhere in any town or country. I was lucky in this, for the strenuous work I put in there built up the physique which all unknowingly was to be so much use to me later in life. My greatest joy was trapeze work, and at a very early age I used to swing high up on a trapeze with a net spread out beneath to catch me should I fall — which was often!

Once a year we held displays. Clad in navy knickers and white sweaters, about a hundred of us, boys and girls, swung Indian Clubs, dumb-bells and barbells to music, followed each other over hobby-horse and then — oh joy! — a chosen few would give a special display of high spring-board jumping, with or without a pole — my very special speciality! Who knows but that this craze for "flying through the air with the greatest of ease" was one of those almost intangible factors which altogether led to my later passion for aviation! To this day, there is nothing which gives me greater joy than a ride on a scenic railway or a steam yacht at the fair, and if I get a chance of a ride on a garden swing, no indignant children can keep me off!

Image 47: Most people in Hull often referred to this building simply as the YPI; but as seen in this picture, its official title was the Hull Young People's Christian & Literary Institute - on George Street. Amy had nothing but praise for this venue, spent much time there and she certainly enjoyed her gymnastics and keeping fit. Initially, Amy was very much a 'team player' – until she reached her two turning points and became a self-imposed hermit, playing truant from school. Courtesy Keith Parker.

Quickly established as the ringleader of my class, I led many a strike and "rebellion". The most famous was the "Revolt of the Straw-hat Brigade". I loathed and detested the ugly straw "bangers" we had to wear and, as a rival school had already adopted the Panama hat (my sister attended this school to my jealous annoyance!), I determined that our school should follow suit. Alas for my youthful ideas of leadership — my prestige was doomed to an ignominious fall. Somehow or other I persuaded my mother to buy me a Panama hat and to put around it my Boulevard ribbon and badge, and one fine morning I turned up wearing this hat, expecting to be supported by the whole class who had promised faithfully to do the same. I was the only one to turn up in a Panama! Punishments at our school were meted out on a most peculiar system, reminiscent to some slight extent of the methods of the police in our modern Russia when dealing with political prisoners. I am told on good authority that, in order to prevent them appearing as heroes to the populace, or even of being allowed to face death with any semblance of heroic dignity, they are shot in their underclothes. Nothing of that kind exactly happened at school, but the method of punishment seemed specially calculated to strip one of all the trappings of a heroine. After the last little episode, for example, I was ordered to stand outside the headmistress's door in my Panama hat until she should come out and see me. The corridor outside her room being a very busy one, almost everyone in the school had the chance to pass me and twit me for being in such an undignified position. There was no knowing when the headmistress might come out, and often I would stand for hours, shifting from one foot to the other, wondering what would happen when the headmistress did appear and my crimes were confessed.

Another curious custom, which has been productive of one of my most pernicious habits, unpunctuality, was to make us stand in the hall whenever we came late in the mornings until we could offer a really acceptable excuse. The result was that I became a most fertile liar without my unpunctual habits receiving any real check. I wonder how many hours I have stood in that hall and how many excuses I have invented! The passing of the years has not in any way dulled this ability!

School days ran on uneventfully enough, their even tenor hardly ruffled by the war days, of which my most vivid memories are of the white bread obtained specially for my father after his illness, which was so serious that, A.I before he was taken ill, he was certified B.3 even at the end of the war; of the ugly purple dresses we sewed at school for little Belgian refugees; of the "Busy Bees" at home, at which I carefully made pincushions out of Union Jacks and started to knit a vest which — fortunately, as my sister said — I never finished; of the air-raids on our town which, to the children, were grand fun as we could stay up and play games and have hot chocolate in the early hours of the morning when we ought to be in bed. Two of the highlights of the air-raids are of my father chasing me when I escaped outside, because I wanted to see the Zeppelins and, as I ran hastily indoors again, my sister — frightened at the noise — collapsed into the coal bucket in the darkness of the cellar in which the family were sheltering. I shall never forget the scare this clattering gave to us all, with nerves strained as they were to breaking point after a raid every night for a week. We were much too young to realize the danger, but how right was my father to be careful. That same night several bombs were dropped in our street, but we all escaped harm. The other incident was when Father decided we should be safer out in the fields, so, it being a wet night, I put on my new shiny white mackintosh. No sooner did I emerge from the front door than I was hailed by a Special Constable with "Put out that light!"

Yes, the war days left very little mark on my life and, although the popular Press adopted the line that it was this early acquaintance with Zeppelins which first put the thought of flying into my head, I do not for one moment believe that to be true. I was far more likely to have felt an intense hatred of them and their awful power.

Image 48: Here is the hat badge worn by pupils at the Boulevard Municipal Secondary School (BMSS). It was the maroon part of the uniform that Amy detested the most: "Not only did I have to wear this hideous colour, but in addition had to have my long curly hair in two plaits with a maroon bow at the end of each" (p142). Spoken like a true tomboy. Courtesy Old Kingstonians' Association.

A tiny incident which seems to have had a much more far-reaching effect on my life was a cricket ball hitting me full in the face and breaking one of my front teeth. Dentistry in those days not being what it is now, I went through my school days with this broken tooth, and from thence onward spoke with a slight lisp. There was no doubt that my looks were seriously impaired from the point of view of my male class-mates and, from being the ringleader, I quickly adopted a most unfortunate inferiority complex which followed me through the rest of my life and probably had a most important bearing on my future actions and character. The boys made fun of me, and I was so self-conscious that I avoided them. I became introspective and withdrew farther and farther into a protective shell of my own making. Instead of spending so much time on games, I acquired the liking for long cycle rides. Many and many a time I played truant from school, cycling for miles into the country with some sandwiches or biscuits in my pocket, which I had stolen from the larder. I also tried again and again to run away. With what object, I do not know. Probably just that craving for "Escape" which has driven many a young man and woman in the last few years to run away from "they don't quite know what" to find "something of which they have no clear idea". My poor parents must have been sorely tried by my inexplicable conduct, but there it was. Feeling myself an outcast, I sought always the hermit's life, and kept away from school as much as possible.

The big influence in my life just at this time came to be the cinema. Seats were 3d., 6d., 9d., and one shilling. On the screen I was thrilled by the lurid adventures of Pearl White, and soothed by appropriate music banged out on the wheezy, tuneless piano. Once an aeroplane appeared on the screen in a news item and, wildly excited, I sat through the whole programme twice just to see it again. For some strange reason that aeroplane appealed to me enormously. It seemed to offer the chance of escape for which I was always looking. I found that there was an aeroplane factory quite near to Hull, and, time and again, I cycled out to keep watch and try to see something of a plane. But I never had any luck. I never saw one close to the ground for I had neither the experience nor the courage to gate-crash the place.

[NOTE: This was during the First World War and the Brough Aerodrome was a top secret government establishment developing war planes during the conflict with Germany.]

Then, one never-to-be-forgotten week-end, when the war was over and hundreds of pilots and planes were out of a job and started that queer era of joy-riding all over the country, an old war-time plane came to a field outside Hull and offered flights at five shillings a time. My sister and I pooled our resources — fortunately I had just had a birthday and was very affluent — and we went up for a flip. Oh, the disappointment! I don't know for what I had hoped. Maybe I thought it would be ten times as exciting as a swing-boat at the fair, but it was not. There was no sensation. Just a lot of wind, smell of burnt oil and escaping petrol. My hair was blown into a tangled mass which could not be combed out for days and I was almost — not quite — cured of flying for ever. That was my first and last experience with aeroplanes until many years later.

Image 49: Actress Pearl White (1916) was one of Amy's Hollywood heroines of the silent movies. Pearl was a woman of action who performed her own stunts. Smith (1967, 26) wrote that "Will Johnson's diary too confirms that she [Amy] was a constant cinema-goer (at fourteen she saw mostly adventure films and then, as she grew older, spectacular historical romances)". Courtesy Commons Wikimedia

In the meantime I was still at school and my future was beginning to loom as the most important subject to be discussed at home. We all sat in solemn conclave to try to solve this difficult problem. I refused mother's offer to keep me at home and train me up to be an efficient and capable housewife. I felt my life needed a larger canvas and I longed to travel and see the world. So far I had never been outside Yorkshire, and London was only a strange and exciting name. I had not the slightest idea what to do with my life and I had no particular leaning or qualification for any one thing more than another. At school I was still enjoying an entirely false reputation for brilliance in my work, and had passed with flying colours the usual matriculation, Oxford Senior and Cambridge Senior examinations. The reason that I took so many was that I stayed at school until I was nearly nineteen and was for two years in the top form trying to grow my hair! And thereby hangs a story.

Our school uniform was maroon, a colour at that time I heartily disliked, but which to-day is one of my favourites. Not only did I have to wear this hideous colour, but in addition had to have my long curly hair in two plaits with a maroon bow at the end of each. An early act of rebellion had been to cut off my hair and fasten back what was left with a hair slide. What triumph it was! No more plaits or ribbons were possible.

My mother was broken-hearted, for she had spent hours and hours brushing my hair and curling it over her fingers and she had just cause to be proud of the result. Thoughtless and selfish, I caused her still more grief and worry. I am afraid I was never a model daughter, and it is only now perhaps, looking back on that period of my life, I realize just how headstrong and wayward I was. Mother must have been an angel of patience. My father, stern and just, decreed as my punishment that I must stay at school until I had grown my hair again. "Bobbing" had never been heard of in those days and, quite rightly for those times, my father considered that a properly brought-up, grown-up girl must be able to put her hair up. So that is why I had so much time to pass so many examinations [Appendix 15C].

At last my future career was decided. Owing to my alleged great cleverness at book-learning, it was settled that I should go to a university, take a degree and be a school-teacher. This seemed as good a career as any other, and I certainly very much wanted a university life. So accordingly, just after my nineteenth birthday, I left for Sheffield, taking with me clothes, books, my hockey stick and violin. The only trouble with my violin was that I could not play. I was learning. You will not be surprised to hear that, after only a few weeks in "digs" with two other students, I and my violin left and sought peace and quiet elsewhere. It was impossible to practise with two noisy girls gossiping and criticizing. My music master in Hull had had high hopes of me as a great violinist. He already foresaw me swaying immense crowds at the Albert Hall with my "art", whilst he would hover in the background murmuring with concealed pride: "My pupil." But this was not to be. Shortly after I went to Sheffield, I caught my finger in the hinge of a car and had it badly squashed. This effectively stopped my practising and I never bothered about my violin again. No, I was not destined to be a great artist, and again I was a disappointment to my mother who, herself very musical and a wonderful performer on organ and piano, had hoped that I might follow in her footsteps. Now I play a gramophone with great skill and that is about all.

Sheffield University had many shocks in store for me — and for my long-suffering parents.

Just nineteen, I arrived there absolutely raw and green, burning to see the world and bursting with my own importance. The first shock came when I sat for the entrance examination. This was more or less a formality to see if you were up to the standard required in the particular subjects you had elected to take for your degree. It also decided whether you were good enough to be allowed to study for an Honours Degree. I failed in everything. I was not up to standard in one single subject. Crestfallen and worried lest I should have to go back to Hull, I went to see the Dean and begged for another chance. Human, understanding and patient, he talked to me and eventually said that I could stay on and take a special subject for my degree. He gave me the choice of Economics or

Logic. I chose the former as seeming far less learned and academic than the latter. I have never regretted that choice, for I learned the essentials of everyday life and have ever since considered that it is a subject which ought to be included in the curriculum of every school.

The next shock came from my room-mates. At school I had always enjoyed an easy popularity and naturally took the lead in everything. Here I found that my careless assumption of superiority made me most unpopular and from the beginning it was evident that we were not going to get on. The other two girls were strangers to each other before we all three came together in these "approved digs". Actually we should have been living in the university hostel but it was full up and we had to await our turn.

Within a few short weeks I had decided to run a lone course, so I gave in my notice and chose other lodgings. Still dissatisfied, in a few weeks I changed again, and yet again and again. In my three years at Sheffield I changed probably thirty times, ending up in a little cottage in Hathersage in the Peak District. Here I was happy. I travelled up to Sheffield, by train as seldom as possible, attended the minimum of lectures and enjoyed the healthy open-air life in the solitude and majestic beauty of the Derbyshire hills. The only drawback to this cottage was the beetles. Never shall I forget them. They are my pet aversion and I have a mad, blind horror of the beastly things. It was sheer ignorance of their obnoxious habits that made it possible for me to continue living there. I laboured under the comforting delusion that they were tied to the ground, that they could not climb or fly. At night-times, lying shivering in bed, I would hear stealthy rustlings behind the walls, but I told myself sternly that this noise was caused by mice, of which I am very fond. One evening as I lit the gas in my bedroom, I saw something black run quickly up the wall and escape inside a crack of which there were hundreds in the walls and ceiling. I held my breath, firmly telling myself that this could not be a beetle because it could not climb. Inspired by a sudden idea, I ran downstairs and brought back a roll of adhesive tape which I stuck over all the cracks. I could not bear to go to bed until every single crack was covered. I was terrified the "black thing" would come out.

Another day, I pulled a book out from my cupboard. To my horror I saw what looked like a long beetle's leg sticking out from one of the pages. Without stopping to think how unreasonable was this fear, I seized the poker and gingerly opened the book ready to stab hard and quickly when the expected beetle should escape. Fortunately no one was there to see the dénouement. The "beetle" turned out to be a pressed fern!

Image 50: When Amy reflected upon her three years at Sheffield University, she stated "I have to admit to myself that they were relatively unimportant and had little bearing on my future life and career, or on my character and temperament" (145-46) – unlike her formative years growing up in the Hessle Road Fishing Community. Courtesy Sheffield Local History Archives.

I could write volumes on my university life, but actually when I look back on those three years and try to see them in proper perspective, I have to admit to myself that they were relatively unimportant and had little bearing on my future life and career, or on my character and temperament. The time slipped past very quickly and, more or less, pleasantly. I did all the usual things, living a life of dances, games, picnics, and parties, interspersed with occasional lectures. Examinations were passed by the barest margin, and only then by dint of "cramming" for the week beforehand. My memory was unreliable, and immediately I had transferred the burden from my overloaded brain on to the paper, I promptly proceeded to forget everything I had so hastily learned. I pawned my books, flirted with the male students, went dance-mad, changed and again changed my "digs", and in short was a typical student of the more unsatisfactory type. By the end of the third year, I was becoming so restless and tired of the life, I begged my father to let me off the fourth year which should have been spent obtaining my teacher's diploma.

Patient and wise as ever, my father quickly saw that I should do no good by remaining, so he gave me permission to leave. Conscience awakened by his kindness and consideration, I decided that I would immediately find a job and so relieve him of the continued burden of having me on his hands. Not qualified now for a teaching job, I racked my brains for a solution of the problem of what to do and eventually I decided on a business career. For this I needed business training, so I worried my harassed father still more to send me to a business college.

For three months I slaved painstakingly at shorthand, typewriting, book-keeping and business methods. Long before I was qualified for a job, I left the college and started to look for something to do. I was restless and — too proud to admit it — already hating the life I had chosen.

Day after day I spent examining the advertisement columns of the local paper. Letter after letter was sent, setting out a grand total of my qualifications. It was inexplicable to me that, with my education, I was not immediately given a directorship! Lack of experience seemed to be the stumbling-block but later, when I had a deeper knowledge of the world, I came to realize that I should have had a better chance of a job had I neglected to boost my wonderful qualifications. For the kind of job I was willing to take in order to get experience, I was too old and had too grand an education. How could a university graduate be told to make the tea and wash up the tea-things, lick the stamps and generally be at the beck and call of everybody!

Eventually I was given a chance. I was offered a job in a Chartered Accountant's office at the princely wage of £1 a week, out of which I had to pay compulsory health insurance, various contributions to odd clubs and societies; five shillings to my mother to "go towards my keep" and make me feel independent (a nice feeling I have never regretted), and the balance for clothes, fares, amusements and pocket money.

I hated the job. I was terribly unhappy. My shorthand was not up to standard and I did not dare to own up to this. I would sit in my chair dreading the bell. With trembling hand I took down straggly shorthand at a speed far in excess of anything I had done at college, with the result that I was unable to read it back and had to go to the dreaded "presence" to ask for words to fill in the numerous blank spaces. My employer died some years later of apoplexy; I was only surprised it didn't happen sooner!

The rest of my time was spent typing financial statements. I was not allowed to make a mistake as nothing must be altered or even erased. No matter how skilful I tried to be with my typewriter rubber, I was always found out. The result was a passion for accuracy and care — which has probably stood me in good stead many a time since and was one of the most valuable ingredients in the whole of my training.

I was unpopular with the other girls in the office, probably both they and I instinctively feeling I was different from them. For one thing I was older. I had attained the ripe age of twenty-one. For another my university degree stood between us. In those days it seemed so terribly important.

After about three months of this life I had a nervous breakdown, and no wonder. Several weeks' cramming for my Finals had been followed by three months' slaving at the business college. My job finished me. I left and went away for a long holiday [Bournemouth].

During all this time I had had a dreadful "secret" on my mind, something I had not dared to tell my parents, and which was worrying me to the exclusion of all other thoughts. Whilst at "Varsity" I had got into debt, and to clear myself had borrowed £50 from a friend [Hans Arregger – her Swiss lover, but not stated here]. Fifty pounds seemed to me an enormous sum and I had no idea how I should ever be able to repay it. During my holiday I made a resolution that I would earn the money to pay back that debt, if it was the last thing I ever did. I searched round for a new job, but decided that I would never go back to that accountant's office, or any other accountants for that matter.

One day, idly turning over the pages of a magazine, from time to time glancing at the advertisements, I found myself criticizing their wording and imagining different lay-outs and ideas. Partly to amuse myself, for my recovery was a long and tedious affair, I rewrote and re-planned advertisement after advertisement. This quickly became an absorbing hobby, and finally was born the idea of taking up the advertising business as a profession. No sooner had I thought of it than I longed to put the scheme into execution. I had no training, and did not know the ropes, but I began by writing to the chief Advertising Agencies in my home town of Hull. I did not ask for a job but for the opportunity to learn the business, and I secured an interview with the leading agency. Here I found that one of the partners needed a secretary, and I was able to make an arrangement whereby I was to give my services as Secretary in exchange for learning the art of writing advertising "copy" and planning advertising "campaigns". In addition, I was to receive a nominal salary of thirty shillings a week. This seemed an admirable arrangement and I started work very soon afterwards, full of energy and enthusiasm.

Alas for my high hopes! In no time at all I found my principal task was sending out thousands and thousands of advertising folders, addressing envelopes, sticking on stamps, and turning in the flaps of the envelopes. Sometimes the whole office worked until ten at night doing this heart-breaking work and I came to dread the days when these folders were due for despatch.

My life in the office was almost unbearable. The other girls were very jealous of my imagined privileges, and I was given the worst typewriter, desk and chair and made to do all the odd jobs in the office. How I loathed it all! And as for learning the advertising profession, what a hope I had! By degrees, however, my persistence in writing "copy" was rewarded and I was given a few minor "accounts" for myself.

I stuck this for a whole year, my principal reason being that I was paying off my debt at the rate of £1 a week, and the joy of seeing this apparently enormous sum gradually diminish to less alarming proportions was worth the effort involved. On the day the last pound was paid I left.

In the meantime, life at home [Park Avenue] was becoming more and more difficult. A boy-and-girl friendship, strongly disapproved of by my parents, had ripened into my first love affair. I could not take any serious interest in a career because I imagined at that time that my career was to be a home and children. I was taking this affair so seriously that I went to cookery classes in the evenings, and learned all I could at home of the art of housewifery. My thoughts and ambitions were centred on only one object — the young man with whom I thought myself madly in love.

My people disapproved for many reasons and subsequent history has proved them to be right. The young man finally married someone else and settled down into a rut which would sooner or later have driven me crazy. My parents knew me better than I knew myself and I have never ceased to be grateful to them for this sympathetic understanding.

However, I was headstrong and was so certain I knew my own mind that I threw over a marvellous chance which my father gave me of seeing the world. When I left the advertising agency, he offered to pay my fare out to Canada, provide me with an outfit and arrange for me to go to an uncle in Canada who would get me a job. How foolish I was to throw away a chance like that, yet that is exactly what I did. I could not bear the idea of such a definite break, so I told my father, half regretfully, that I did not want to go. Poor father, what a lot of worry and trouble I gave him!

The next move was to get another job. The thought of Canada and the wide world had, probably quite unconsciously, made me restless and I was no longer satisfied to stay in Hull. I made the big decision to try my luck in London, and, with an allowance from my father enough to last me for a month, I set off full of hope.

In spite of my increase of years and added experience, I was still almost as green as when I first went to Sheffield, and London completely took my breath away. Everything was so rushed, noisy and expensive and I felt lost and lonely. I knew no one, but my father knew what he was about when he decided to throw me on the world to sink or swim by my own efforts. I really believe he had more faith in me then than I had in myself.

My adventures in London would fill another volume, but again I want only to select those incidents which proved to be the shadows of coming events.

So far as finding lodgings was concerned, history repeated itself. Restless and dissatisfied, I changed continually, living in women's hostels when funds were low and in furnished "bed-sits" when jobs materialized.

The first job was selling ribbons in a large store. My original idea when coming to London was to continue my advertising career, but no one seemed to share my own opinion of my capabilities. I shall never forget seeing an article in a newspaper by Sir Charles Higham, one of the best-known advertising experts. He was deploring the lack of talent and enthusiasm in the younger generation, so I seized the opportunity to write to him and send my portfolio of "copy" and "lay-outs". His reply was courteous but firm. I was not sufficiently far advanced for a job in his firm, but I could write to him again. When I met him at a dinner many years later, I reminded him of this little episode. He laughingly said, "Well, I told you to write to me again later. I'll give you a job now."

After spending my allowance on fruitless weeks of searching for an advertising job, I decided to try my luck with a new scheme which was being operated by a certain well-known West End store, whereby University Graduates, and Ladies of Gentle Birth and Breeding, were promised a minimum salary of £3 per week and training, in return for their services. It was, of course, hoped that such people would be able to introduce new customers to the store. Unfortunately for them, they had no idea just how alone in London I was, or I should never have got the job.

At nine o'clock on a certain Monday morning I took my place behind the counter to sell chiffons, silks, satins and ribbons. I loved the delicate shining materials and was quite prepared to throw my whole energy and enthusiasm into making a success of this new job. The prospect of meeting new people, of persuading them to buy, and wrapping up parcels and handing over change — reminiscent of my childhood games — seemed sufficiently exciting.

At the end of the most tiring week of my life, I had no illusions left. Being the junior and last-comer of the department's five assistants, I was only allowed to serve when all the rest were already busy — which never happened. I spent the whole of every day on my feet, from nine o'clock to six, unutterably bored and tired, for nothing is worse than inactivity. The only jobs found for me were tidying away the boxes rifled by senior assistants. At the end of the first week I received in an envelope my week's salary of £3, according to the agreement, along with a slip of paper stating quite crudely that my services for the week were valued at 4s. 7½d.

The second week was an exact repetition of the first. At the end of the third week I was taken ill with influenza which kept me away for the fourth and last week of my month's trial. I decided not to go back.

Life now seemed incredibly hard. I was ill, friendless and almost out of funds. Pride would not let me write home for help, and, in any event, I still had plenty of faith in myself, incredible though it may seem as, so far, I had not been a conspicuous success.

About the only person I knew in London was a distant cousin, who came to my rescue just when I badly needed a little encouragement. She tried to get me a job at the "Times Book Club" where she worked. This appealed to me. A desk all my own, cheerful with its vase of flowers, and apparently as many books to read and handle as I could manage. I was unlucky, though, and failed to get the post. She then kindly introduced me to a friend of hers, a solicitor, who had known me as a small child and knew my family well. Taking pity on me, he offered me work in his office, beginning at £3 a week. Full of gratitude I accepted and started work as soon as I was fit enough.

I spent three years at this office and only left to take up a flying career. My luck in getting the job was definitely the turn of the tide. There were still plenty of struggles and difficulties ahead, but my feet were firmly planted on the first rung of the ladder which was to lead me to fame — though little did I know it then.

The work at the office was as congenial as any office work could be to a person of my temperament. Soon I had carved out my own little niche. As I had not stepped into any particular vacancy, I was allowed to find my own work, more or less. I found the law intensely interesting, and began to specialize on company law.

My life outside the office, too, became much happier. I linked up with a girl I had known at Sheffield University [Winifred Irving] and we shared lodgings together, or rather, a variety of lodgings for, in my restlessness, I took her along with me in my constant changes.

We finally came to rest in an attractive little room in a big gloomy-looking house in Maida Vale. Our "garden-room" as we called it, overlooked a big open space with pleasure gardens and tennis-courts. The sky was alive with aeroplanes for Maida Vale was not far from Stag Lane aerodrome. The noise of the engines – hated by the rest of the inhabitants — was music to my ears. Often in the middle of a set of tennis I would stop and gaze wistfully skywards. I envied those pilots. I longed for the freedom and detachment it seemed they must enjoy. Nor did my interest in these aeroplanes vanish when the novelty of seeing them had worn off. I became more and more absorbed. In some intuitive way I felt that there was a link, as yet to be forged, between the planes flying over that Maida Vale house and myself.

The mere longing to fly was not new. I had always, subconsciously, wanted freedom and adventure and I must have felt that flying could give these to me. However, with aeroplanes, for the first time in my life, near enough to me to be seen and heard every day, I found myself thinking more and more about them and I began actively to long and to want to get nearer to them, to see them and, if possible, to fly in them. That I could ever afford to learn to fly myself seemed too wild and impossible a dream and my fears were confirmed when, one day just for fun, I wrote to the de Havilland Aircraft Company to ask for their prospectus. Five guineas an hour for instruction! Yes, I was right. It was impossible, and regretfully I abandoned the idea.

Then one day happened that trivial little incident which was to alter the whole course of my life. One fine Saturday afternoon, drawn irresistibly by those tiny darting planes, I climbed on top of a bus going in the direction of Stag Lane. As I drew nearer, the planes came lower and lower until they seemed to be landing almost on my head as they glided down into the aerodrome. Wild with excitement, I jumped off the bus and ran up Stag Lane, to find at the end a notice "London Aeroplane Club. Private." My keenness to explore outweighed my natural shyness and, disregarding the notice, I walked on until the green of the aerodrome itself came into view. Two yellow aeroplanes were lined up ready to fly and caught my interest. There was a small clubhouse, in front of which were people seated on deckchairs. No one took the slightest notice of me as I nervously approached an unoccupied chair and sat down. Every minute I expected to be challenged. Nothing happened. Interest was centred on the aeroplanes about to take off. For half an hour or so I sat enthralled. I had to learn to fly. I simply had to. It was fast becoming an obsession. At last, summoning up all my courage, I asked the instructor how much it cost to learn to fly. I could hardly believe his reply! "Two pounds an hour for instruction. Thirty shillings an hour solo. Three guineas entrance fee and three guineas subscription. Takes from eight to twelve hours to learn."

I could afford it! With pulses racing with excitement, I quickly added up ways and means. I had £1 in my purse. "Could I start straight away?" No, apparently I could not, my name had to go before the club. I must await my turn to be duly elected. But not even this information could damp me. I could afford it on my salary of £3 a week and I made up my mind there and then that, at whatever sacrifice, I was going to be a pilot. To flying as a subsequent career I never gave a thought. I merely wanted to fly a plane all by myself. For the moment that sufficed.

This was the most important milestone in my life. It was the turning-point, the beginning of Part Two. And it is at this point that I have decided to finish this chapter on "Myself when young". I was then only twenty-three, but the story of learning to fly, with its set-backs, bitter disappointments, heart-breaking struggles and final triumph, is another tale and too long to be told here.

I have tried to link together the chain of incidents leading to my flying career, and whether they answer the two favourite questions of "Why did I learn to fly?" and "Why did I fly to Australia?" I must leave to you. A friend of mine once summed it up very pithily: "Well, you've always been a woman of extremes — Australia and all that!" To me, looking back on all these years, the sequence of events seems natural and inevitable. Born of an adventurous family — my grandfather, a Dane, had run away from home as a small boy, while my father had been one of the first out in the Klondyke gold rush — it was not surprising at all that I had the same kind of urge. Had I been a man I might have explored the Poles, or climbed Mount Everest, but as it was, my spirit found its outlet in the air. Everything in my life since has spelt adventure and I hope always will. As I write this, I am still "young" by the standards of the publishers of this book, who say that "there is no age limit to youth" but add that "thirty-five can be taken generally as the starting-off ground to middle age". My career, I hope, therefore, is only left here to be continued.

THE END

~~~~~~~~~~~~~~~~~~

Image 51: Amy's fortuitous bus ride from Maida Vale to the North London Stag Lane Aerodrome changed the rest of her life. Lady Luck smiled upon her that day. Courtesy Wikimedia Common and de Havilland Advert c.1928.

AG Comments:

From the start, Amy showed how she was born with a "spirit of adventure" in her genetic make-up. She needed to escape from her family home - even as a toddler. Maybe, I should not be too critical of Amy wanting to get away from Hull (or anywhere she might have been born) because it was deep in her DNA. As a three-year old runaway, she was drawn to a nearby house with a "bright green" door – obviously, a favourite colour from early days and perhaps another reason she painted Jason that colour when she made her 'escape' from Britain in May 1930!

Throughout her writing, she made good use of imagery. I like the description of how she loved fairy-tales and a beautiful princess falling in love with a handsome young peasant trespassing in the Palace woods, who turns out to be a prince in disguise.

Lucky black cat china ornaments played a part even in Amy's earliest childhood days – never mind her adult life! An interesting snippet was how she persistently stole the same black cat object from her granny's mantelpiece three times. Talking of 'luck', I did a quick count of this word and Amy used it six times in the above piece (along with 'restless' = five times; and 'escape' = four times).

I find it highly significant that Amy Johnson's first ever encounter with an aeroplane was when she was sitting alone in a Hessle Road cinema watching a movie (c.1916/17). She clearly described her positive reaction as "wildly excited" and "seemed to offer the chance of escape for which I was always looking". This deeply profound effect speaks volumes. Young and impressionable, Amy made an everlasting, heartfelt mental note that an aircraft was her best escape from ugly streets and smelly fish, from silly uniforms and unfaithful friends, from broken teeth and tormenting boys.

Again, Amy's openness is admirable when (as an adult) she boldly declared "I hated the job. I was terribly unhappy". This was primarily because she was not accepted by the office workers, especially the other women. Amy's pattern of being ostracised - in Sheffield, Hull and London, repeated a precedent set at the Boulevard School whereby she ended up becoming a hermit yet again. In a convoluted way, the question arises, was Amy's social isolation self-inflicted? Or was this pattern built into her natural makeup?

Another facet of Amy's personality was that she soon became "bored" by many situations. Rather than endure boredom and social expectations, Amy's escape mode risked her being excluded by others – but that high price was one she was prepared to pay.

Amy shows great affection towards her parents in this chapter, especially her father. She was writing this in 1938, when she was 35, had just come through her divorce with Jim Mollison, and was very dependent upon her family for their love and support through that crisis time.

Regarding Amy's relationship with her parents, practically all biographers highlight how the Johnson's first-born daughter was "high maintenance" (to use a modern-day phrase to describe a difficult teenager). Perhaps what Will and Ciss were unable to perceive in their day was that their intelligent child was rebelling against the set rules and expectations for girls / women – even though Amy herself did not fully understand why or what she was kicking against – nor from what she was escaping.

Young Amy was too much for her parents; but they, likewise, had no understanding or awareness of their brilliant daughter's dilemma with life. She was always questioning the status quo, especially the role of women in society. For Amy, it was natural to challenge what she perceived as unfair or boring.

She hated being boxed-in by others' expectations. She was a clever, imaginative, individualistic woman at odds with the parochial and patriarchal world around her. All she wanted was to escape from conformity. In essence, Myself When Young clearly demonstrates a freethinking, strong-willed and self-determined individual.

Her final paragraph is especially striking. She was keen to highlight her "adventurous family" in that her Danish grandfather had runaway to Britain and her dad had ventured off in search of gold in the Klondyke. Nevertheless, she still avoided even the slightest hint that either of them had anything at all to do with Hull's Fishing Industry! My one wish was that she had mentioned fish!

#################

(15B) At Amy's First School - interview with a Fellow Pupil

Appendix 15B: HULL DAILY MAIL. 9 October 1973. Newspaper. At Amy's First School. Women. Miss Humber. 6. Hull History Centre. Archive news-cutting.

IT IS AGREED by most educationalists that a child's first school years are of vital importance in shaping its personality, and influencing its interests and future learning patterns.

It is somewhat surprising, therefore, that most biographers of Amy Johnson, the most famous woman in Hull's history, have started her life story at the point when, aged 10, she went to the Boulevard School, later Kingston High School.

Before that, she and her sister, Irene, attended a small private school. This was Eversleigh House, where one of their fellow pupils was Janet McDougall, now Mrs Robson of Voases Close, Anlaby.

The school was at the corner of Glencoe-street, now occupied by a pet shop [1973], and later it was moved to the other side of the street, next to a dentist's surgery.

Mrs Robson tells me that headmistress was Miss Ada Knowles, a graduate with a particular interest in geography. She was assisted by her sister Miss Bertha Knowles and by Miss Mollie Burrell and a Miss Rank.

Miss Ada was very particular about the behaviour and good manners of her pupils and would never allow them to venture outside without gloves. Yet the sisters were not too proud to get down on their hands and knees and scrub their little school clean from top to bottom after the children had left for the day, as Janet discovered when she went back one afternoon to collect something she had left in the classroom — a piece of forgetfulness which earned stern disapproval.

Mrs Robson, who attended the school from 1907 to 1912, remembers well the geography lessons taken by the headmistress.

"She would put up a map and point to a place on it. The first person to name the place, river, town or whatever it was, received a mark, and Amy and I used to race for it." She told me.

Was this, she wonders, the beginning of the interest in faraway places, which was to culminate in that all-time heroic solo flight to Australia?

"She was a very high spirited little girl and I remember her dancing on the teacher's desk when she was out of the room," said Mrs Robson.

Image 52: Eversleigh House (now a mini-market) at No.557 Anlaby Road near the corner of Glencoe Street was run by the two Knowles Sisters (Ada and Bertha) plus two assistants. This was where Amy is reported to have been "a very high spirited little girl and I remember her dancing on the teacher's desk when she was out of the room". Copyright Alec Gill.

The Johnson family at that time lived in Alliance-avenue. Many years later, Mrs Robson was to meet Amy's mother again. After the death of her first husband, Mr Alfred Sheppard, Mrs Robson worked as a nursing auxiliary at Castle Hill Hospital, where Mrs Johnson became one of her patients.

She told me that even in the throes of an illness, Mrs Johnson was a wonderful person, keeping the ward alive with her bright conversation and the games she organised.

Mrs Robson is anxious that the little school where Amy received her first lessons should receive some of the credit due to it.

Having received the major part of my own education at a similar type of establishment, I know of the wonderful work these one-woman private schools could sometimes do.

With just a little help with the auxiliary subjects, an arts graduate with a good brain and wide interests could often pass on to her pupils a uniquely balanced and well-co-ordinated appreciation of the world around them and fire them with a love of learning which could lead them anywhere they cared to go.

~~~~~~~~~~~~~~~~~

AG Comments:

Janet Robson's interview with Miss Humber provided a fascinating insight about Amy as a young girl and corroborated Amy's own account of herself when young (Appendix 15A). Janet's two anecdotes about the love of geography and dancing on the teacher's desk are gems. This piece also portrays an elderly Ciss Johnson as a colourful character organising games in a hospital ward.

##################

(15C) New Year's Day in China
by Amy Johnson

Appendix 15C: JOHNSON, Amy. April 1921. Journal. New Year's Day in China. The Boulevardian. 21. 2. 27-28. Available via The Old Kingstonians' Association at the Carnegie Heritage Centre.

IT IS THE PRIVILEGE or bane of an Englishman in China to have his holiday season twice, for the Chinese festivities do not, as in England, centre round Christmas, but round the New Year. The foreigner residing in China may, or may not, take notice of his own holidays as he wishes, but he must recognise those of the Chinese. As the Chinese reckon their year differently from other countries, their New Year's Day varies from year to year, generally occurring, however, in our month of February.

This festival, which not only extends through New Year's Day, but through the week or two following, is used for the purpose of observing several things besides the fact of its being the first day of a New Year.

First, it is a kind of sabbath; a time of rest. The thrifty multitude that has been steadily plodding along month after month, with no diversity or relaxation, at last pauses, and strives for a few brief days to find happiness. Only the very poorest remain at work. The richer the man, the longer is he freed from his customary employment. The Chinaman not only rests during this period, but worships. At home the whole family assembles, and first heaven and earth are worshipped, then the family gods and deceased ancestors, and after that the younger members make their protestations to the older ones. Finally, after a presentation of gifts, the men enter the temple, and amid the sound of the drum, gong, bell, and chant of the priest, burn incense and bow with head to the ground before their favourite idols.

Secondly, this festival is a kind of bonfire day, for many loud and brilliant fireworks are sent up by the priest, and long strings of firecrackers are lighted on the thresholds of the houses, it being believed that all evil spirits will thus be kept away during the ensuing year.

New Year's Day is the pay-day of the Chinese. All debts must be satisfactorily settled before the old year has gone. If perchance any have failed to settle up their accounts with their creditors, they will be found the next morning hastening up and down the streets with lighted lanterns to show that the work of the past day has not been finished. If the debtor has nothing wherewith to pay, the creditor may enter his house, and take or destroy to his heart's content. Debts to the gods are also settled by the more devout Chinese by means of gifts and prayers.

This festival is also the birthday anniversary of all the Chinese. If a child is born only a few days before the close of the one year, at the beginning of the next he is reckoned as two years old, the reason being that he has seen two years. Every person is therefore one year older when the new year begins. In growing older the Chinese always seem to rejoice.

Image 53: Amy's History Notes when in Class III D. I am not sure where this particular document is held, but a good set of The Boulevardian Magazines are stored at Hull's Carnegie Heritage Centre under the care and control of the Old Kingstonians' Association. Courtesy Hull: The Good Old Days - Keith Parker.

There are three other customs which are observed at the New Year's Festival, one being that the memories of all newly-deceased members of a household are honoured by posting up of mottoes written on blue or white paper, either colour being a sign of mourning. Another is that the period is introduced by a general wash-day. Every person has a bath, and this event is all the more important in the lives of a few, from its occurrence only once a year. After cleansing, the Chinaman attires himself in his most gorgeous robes and then marches forth to pay calls. About the third day the women begin to exchange calls, likewise showing a desire for fine dress, elegance and flattery.

This season, last of all, is one of amusement. Rich men hire actors, build stages in the open-air, and amuse the masses. High officials, who close their offices for thirty days, give to the poor, hire theatres, listen to plays or receive guests to well-prepared feasts. A large portion of the people gamble, drink, or smoke, and so frequently quarrel, thus marring what would otherwise be a period of rest and enjoyment.

A. J. (VIA)

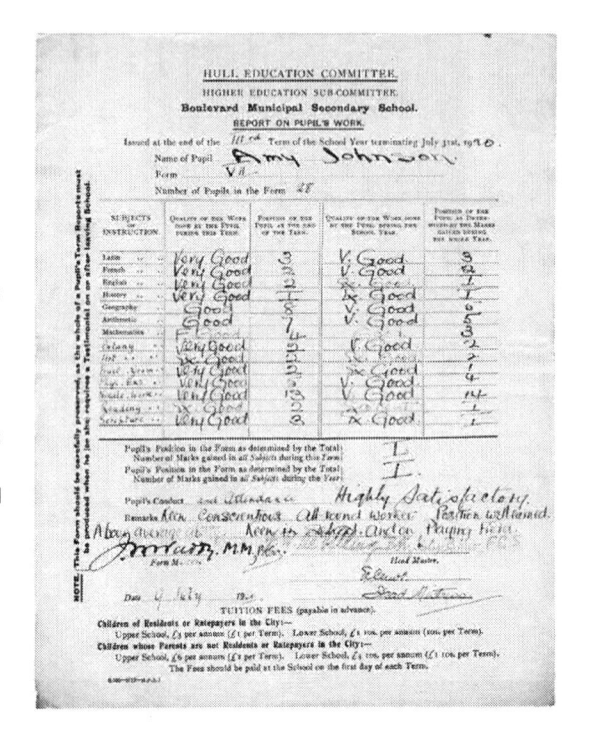

~~~~~~~~~~~~~~~~

Image 54: Here is Amy's School Report for 9 July 1920 – just turned 17 years old. She seemed to have done very well. I am not too sure, but it looks as if she was First in her class that year? The Tuition Fees at the bottom of this form suggested that parents had to pay for their children to attend. Courtesy Christine Pinder and Rob Wilson.

AG Comments:

This essay might have originated from a set piece of work - by a teacher at the Boulevard School – and reproduced in the magazine because it was so well written. We might never know. I have highlighted in my book how Amy was very superstitious. Therefore, I see this topic as an early indication of Amy being drawn towards the local folklore beliefs of the Hessle Road culture that was all around her. It was perhaps another form of her schoolgirl counter-reaction against her parents' Methodism.

It seems a mature piece of descriptive writing which has been well researched - with some good insights into Chinese New Year rituals. I wonder where and how Amy conducted her research for this article.

~~~~~~~~~~~~~~~~

NOTE - a point of interest regarding Amy's second attempt at a record-breaking flight in 1931:
After Amy completed her flight to Australia (May 1930) and became a global celebrity, she was soon eager to take off again on another record-breaking venture (perhaps she continued to 'get bored' even when famous?). Where did she choose?

Well, it could be argued – in a convoluted way - that her above Boulevardian article had some sort of double influence.

Where: her choice was to fly to China.

When: New Year's Day 1931. Coincidence? Who knows?

If this was taken by Amy as a lucky omen, she was grossly wrong. It was a bad time and place to undertake any flight. She seemed to have overlooked the severe winter weather conditions she was bound to encounter as she flew over Eastern Europe heading for Siberia. She took off from Croydon on Thursday 1st January 1931 at the controls of Jason III (G-ABDV). Dense fog over Poland soon brought her plans to an abrupt halt with a forced landing in a forest clearing. Subsequently, she got a train to Moscow. The Soviets made a big fuss of Amy before she returned to the UK. On the 28 July 1931 Amy, with Jack Humphreys, took off from London for Tokyo aboard Jason II (G-AAZV) – in better weather conditions.

NOTE: Once again, I am thankful to Peter Nicholson - of the famous Hessle Road Furniture Shop Marsh Nicholson for originally loaning me a copy of this Amy Johnson article back in November 1992. His father George attended The Boulevard School at the same time as Amy Johnson.

I am also grateful to Christine Pinder and Rob Wilson for their help with material about Amy Johnson from the archives of The Old Kingstonians' Association located at the Carnegie Heritage Centre.

##################

(15D) Sky Roads of the World
by Amy Johnson

Appendix 15D: JOHNSON, Amy. 1939. Book. Sky Roads of the World. London. W & R Chambers.

DEDICATION:

This book is dedicated to all those who fell by the airwayside, for nothing is wasted, and every apparent failure is but a challenge to others.

PREFACE:

In this preface, I wish merely to explain that this book is meant to be rather a romantic story of the world's great sky routes, as seen by a pilot who has flown over most of them, than an encyclopaedic history or an information bureau - although the facts and figures quoted are as accurate as careful investigation can make them.

I have tried to give life to the network of air-lines which cover any map of the world of to-day; to make you see the countries and oceans over which they pass, and appreciate some of the human drama and labour which have gone into their being.

I have written mostly of people I know and places I have seen, but, above all, I have tried to recreate a breath of the magic of the air which, alas! we so often forget to notice in this present hard commercial age.

AMY JOHNSON

CONTENTS:

I. THE DAWN OF THE AIR AGE - 9
II. TO AUSTRALIA AND NEW ZEALAND - 31
III. TO CAPE TOWN VIA CAIRO - 59
IV. ACROSS THE SAHARA TO SOUTH AFRICA - 81
V. LINKING THE NEW WORLD WITH THE OLD - 101
VI. THE CONQUEST OF THE PACIFIC - 147
VII. AIRWAYS OF THE BRITISH EMPIRE - 179
VIII. AIRWAYS IN EUROPE, ASIA, AND THE FAR EAST - 233
IX. THE THREE AMERICAS - 260
X. GLIMPSE INTO THE FUTURE - 284

Chapter I: THE DAWN OF THE AIR AGE (extract only)

It is generally admitted as an axiom that civilisation develops in proportion as transport facilities improve. Countries which to-day still use donkey transport occupy the same status in the world that we ourselves occupied over a thousand years ago. Fast transport means facilities for trading, for travelling and getting to know our neighbours, for the interchange of ideas and customs, for the exchange of goods, raw materials, and foodstuffs. In this hurrying modern world it is very evident that the slow-thinking, slow-acting countries will be left behind in the race for power and precedence.

We are living to-day in the Air Age, and ours is one of the leading nations in our present civilisation because we have realised - slowly, but none the less surely - that to neglect this new, swift, vital means of transport would be the first quick step to our downfall.

For centuries man has felt an urge to fly, without quite knowing why. Certain it is that he cannot have visualised flying as a means of commercial transport, carrying loads of passengers and goods at incredible speeds over all the wide earth. It is far more likely that he merely envied the birds their freedom and wanted to be able, like them, to shake off the fetters of gravity and soar far afield, adventuring into space.

The earliest attempts to fly were all based on efforts to copy the movements of birds. Flapping wings, and even feathers, played their part in the very earliest contraptions, but never was man able to fly like a bird, and his struggles led only to failure and death.

Balloons, of course, flew more than a century ago, but the principle was simple, and the mere act of rising into the air with a hot-air balloon was no extraordinary achievement. The harder part, indeed, was to come down again! Nor could balloons, however successful at going up and coming down, ever be of any commercial use, drifted hither and thither as they were at the mercy of the elements.

What was wanted was a machine that would take off by its own power within a reasonable distance, fly straight to a predetermined destination, and glide down to land safely, also in a reasonable distance.

At the end of the nineteenth and the beginning of the twentieth century, many active brains were puzzling over the problems of flight. A German, Otto Lilienthal, made a glider which flew through the air and glided down to land, but, without an engine, it was powerless to take itself off or to stay up. He then built an engine and fitted it into his glider, but he had tried the Fates too far, and they mercilessly sent him crashing to his death on his first trial flight.

About the same time were a few enthusiastic Englishmen working on similar lines, and two or three Americans. Each had his own ideas, and it was a curious coincidence that experiments reached approximately the same degree of success in more than one country almost simultaneously, although everyone was working independently. There is, of course, only one theory of flight, and the laws of aero-dynamics must follow their marked-out path.

One of these inventors, however, was bound actually to fly before the rest, and it was to the Wright Brothers, working secretly in a remote part of America, that the honour finally fell. They will go down in history as the men who first flew in a heavier-than-air machine. It is significant that these two, Wilbur and Orville, are always referred to as 'the Wright Brothers', and are given equal credit for their great achievement. They worked together in close harmony, and one did as much as the other to attain success. Actually it was Wilbur who, on a bitterly cold day, 17th December 1903, really and truly flew. His flight lasted only twelve seconds, and was followed by a slightly longer one by Orville. Taking it in turns, they made two more flights, the fourth one lasting 59 seconds, covering a distance of 852 feet over the ground against a twenty-mile-an-hour wind, their speed a bare 36 m.p.h. After the fourth flight, a gust of wind blew the plane over, and it had to be dismantled and taken home.

Improving on their original design, they continued their experiments until, two years later, they were able to cover a distance of 24¼ miles.

Meanwhile, the news that the Wrights had flown had filtered through to other countries, and was barely believed. Their experiments had been carried out mostly at Kitty Hawk, a place far from reporters and crowds. It was not so much secrecy that they desired as to be left in peace, but this made it difficult for them to prove their claim that they were the first men in the world to fly. Long and painful was the litigation that necessarily followed their efforts to prove it and to patent their designs. America, their own country, refused to recognise their claim, and it was Great Britain that finally came forward to give them their due.

Eventually, Orville Wright agreed to sell his patents to the British Government for the nominal sum of £15,000. They also bought the original plane, and it was given an honoured place in the South Kensington Science Museum, where it still is to-day. Meantime, Wilbur Wright had died of typhoid in 1912, worn out with the anxiety and distress the litigation had caused him.

It seems incredible that in the short space of thirty-five years the aeroplane has made such progress that speeds have gone up from 35 to nearly 500 m.p.h., distances covered non-stop from a few hundred yards to over 7000 miles, and heights from a few feet to the stratosphere, more than ten miles high. Most civilised countries are now a network of airlines, practically every town has its airport, and there are ever fewer people who still will own [up] that they dare not fly.

Orville Wright I met in the States some four years ago. I gazed with awe and respect at this quiet, retiring, grey-haired man who was in truth the first conqueror of the air...

++++++++++++++++++

Chapter II: TO AUSTRALIA AND NEW ZEALAND (extract only)

After this comes my own flight, which I started - 5th May 1930, arriving in Australia 19½ days later. From this flight I will give you just a few of the high-lights to show what kind of difficulties we were up against in those days. Why I wanted to fly this particular route I really do not know, except that it was the greatest distance I could fly a Gipsy Moth. Anything beyond that involved sea crossings for which I had not the range. Moreover, perhaps the route had a certain fascination, a glamour that attracted me. I was told hair-raising stories of torture-loving bandits in the mountain fastnesses of Turkey and Iraq, of wild beasts in the desert and jungle, of cannibals in the farther islands of the East Indies, and of sharks in the Timor Sea. My vivid, almost school-girl imagination was fired. Fed as I had been since childhood on fairy-tales, stories of the Arabian Nights, Greek and Northern Mythology, and innumerable books of adventure of the Rider Haggard and Jules Verne type, it was no wonder that the dangers frightened but enchanted me. Difficulties (technical, financial, and parental) had no power to stop me once my mind was made up, and I rode somewhat rough-shod over bunches of red tape, with the help and encouragement of Sir Sefton Brancker, then Director of Civil Aviation and responsible for obtaining Lord Wakefield's invaluable financial help for the flight.

So far as preparations went, I was faced by a complete and utter ignorance on the part of myself and everybody else. The first person I found to give me any advice at all of a helpful nature was Captain Hope, from whom I bought my machine, a second-hand Gipsy Moth which had already flown some 35,000 miles, mostly carrying pictures which Hope was bringing home for the newspapers. It was he who told me what sort of equipment I should need, where to go for maps, and added some valuable information regarding aerodromes, weather, etc.

Besides Captain Hope, the only other persons who were able to help me were Jack Humphreys, the chief engineer at the London Aeroplane Club, and Mitchell, his able second-in-command. Together they put me through my engineer's licences and taught me all I ever knew about an engine. They advised me what spares to take, and what to do when things went wrong with the engine or plane; but for their training I could never have got through.

For maps I had to take whatever I could obtain - good, some bad. Courses and distances were marked out for me, all available aerodromes, landing grounds, and race-courses indicated, and the maps cut into strips and put on to rather clumsy 'rollers' so that I could unroll them bit by bit as I continued my course. (In theory ideal, but in practice I found they would roll backwards!) Once in possession of a list of aerodromes, I set to work to plan my route. I laid out the routes taken by previous fliers and decided on at least one innovation. It seemed to me a sheer waste of time to go a long way round via Lyons, Rome, Cairo, Damascus, and so on to the Persian Gulf, the route by Ross Smith; or by way of Rome, Malta, Ramleh [Palestine], that chosen by Hinkler. From the map it appeared quite obvious that any time I could save must be on the first part of the route, as, after the Persian Gulf, one was almost bound to go the stereotyped route taken by the rest, across India, through the Malay Peninsula, and through Dutch East Indies to Port Darwin. I drew a line from London to Basra on the curve of the Gulf, and discovered it took me through about a dozen European countries, right across the dreaded Taurus Mountains (about which such terrible stories were told by Air Force pilots of being held to ransom by bandits if captured), and through the Arabian Desert to Aleppo and Baghdad. The prospect did not frighten me, because I was so appallingly ignorant that I never realised in the least what I had taken on, in spite of what I was told and all the awful warnings I received. I decided to take this route. I was not concerned with political or commercial questions, and only thought of saving time so that I could beat Hinkler's record.

The first thing I was up against was permits. I found I had to give six weeks' notice and a deposit of £50 to the Air Ministry to cover expenses of cables, etc. I was also told that it was most unlikely Turkey would give me permission, and as an actual fact this permit had not arrived when I finally left on my flight. The assorted array of documents I eventually collected was something amazing to behold, including a message to the bandits asking them to guard me safely, as ransom would be paid. (I do not remember how much was promised. Certainly, however, much more than I was worth at the beginning of my flight!)

The next thing was arranging for supplies of petrol and oil. Most of it had to be shipped out specially, although on Air Force aerodromes supplies were available. In some places were unused casks from other flights.

Equipment of the machine was easy, as I had had the good fortune to buy a plane with long-distance tanks already fitted. These gave me a range of 1150 miles, or 13.5 hours' flying in still air at my cruising speed of 85 m.p.h. For instruments I had an airspeed indicator, an altimeter, a turn-and-bank indicator, and one single compass. In a modern long-range aeroplane there are almost a hundred dials, knobs, and throttles! The front cockpit (my Gipsy Moth was a two-seater open biplane with one hundred-horse-power engine) was covered over, and into it I crammed a medley of tools and kit reminding me of a village store. Besides a goodly supply of tools and spares there were tyres, inner tubes, clothes, sun-helmet, mosquito-net, cooking-stove, billy-cans, synthetic fuel, flints, revolver, medicines, first-aid kit, air-cushion, and Heaven knows what else. Everything, in short, that might come in useful in the event of a forced landing, and I had to legislate for deserts, jungles or seas, mountain-tops or swamps, heat or cold, day or night. Fastened with rope to the side of the machine was a spare propeller, on the seat of my cockpit was a parachute, and in every available corner were emergency provisions. It was the only way - like taking out an umbrella every day so that it will not rain.

It took me nearly six months to prepare for the flight, and even then I left without my Turkish permit, and with last-minute news that my supplies of petrol for Timor would only be through in another two months. Time was getting short. No one knew very much about the weather conditions I should find, and when I went to the Air Ministry to make inquiries I was shown into the room where records were kept and told to help myself to any information I could discover in them. I had a vague idea that monsoons occurred during the summer months, and I read somewhere that the monsoon storms broke in the middle of May - but I never found out where (until I started!).

The day I actually set off - 5th May 1930 - I was aiming for Vienna, 800 miles away, and available reports gave me the weather only as far as Paris. After that I had to take what came. At Vienna there was an excellent aerodrome, with mechanics to do all the work for me, and except for the inconveniences of pumping all my petrol by means of an old-fashioned hand-pump, sitting in the discomfort of an open machine on a hard parachute for ten hours, instead of the four in which this 'hop' could be done nowadays, being without weather reports, and knowing the whole time that if anything went wrong with my engine - more than likely – I must come down where I could, this first day's flight did not differ greatly from what it might today.

It was at Constantinople that I had my first taste of such ignorance on the part of several willing enough helpers that, when lifting the tail of my plane to wheel her into the hangar, they did so with far too much gusto and landed her on her nose! Fortunately, no damage was done.

I have often had arguments as to which is the greater strain - to fly through a long period at slow speeds, stopping for a night's rest every now and then, or to rush straight through, flying day and night practically without sleep. I have had both experiences and unhesitatingly say that the former is more tiring by far – at any rate, for my temperament. On my first Australian flight, after taking more than twice as long to cover any distance as it would take to-day, in an open cockpit, which is fatiguing because of the noise (we always had stub exhausts to try to increase speed and also to enable us to see the accuracy of the petrol mixture at a glance) and exposure to the weather, I had spent hours on the ground looking after my engine, because at very few places were there trained mechanics, and it was vital to carry through a certain amount of maintenance work at the end of every day's flight. For 19½ days I had an average of only three hours' sleep per night, whilst on my last flight to the Cape I had only three hours' sleep during three and a half days. On the last Cape flight I never touched my engine except to fill with petrol and change the oil occasionally, whereas on the flight to Australia I did at least three hours' work at the end of every ten hours' flying.

In Turkey I was held up for hours owing to my permit not having been received, and it was only after a great deal of difficulty that I was allowed to proceed. Once away, I met storms and bad weather in the Taurus Mountains, of which I had no knowledge beforehand, as I never had another weather report except at the better-equipped Air Force aerodromes, and even then merely local conditions were given.

Persia demanded a health certificate, and only let me proceed without one when I pointed out that I should be out of their country in a few hours without again touching ground, and therefore I had not much chance to spread any strange and wonderful diseases, supposing I had them. They had never thought of that point of view, and let me go.

Where aerodromes did not exist I had to land on race-courses or prepared grounds. At Rangoon, searching for the race-course in a tropical downpour, I mistook for it the playing fields of an Engineering College at Insein, just outside the city. Landing in a space far too small for my plane, I ran into the ditch and did a considerable amount of damage. In this country it would have meant major repairs - new wings, blue prints by the bundle, weeks of hard work, and quite a lot of money. Having none of these at Insein, we set to work to do the best we could. Wings were patched with men's linen shirts; new ribs, bolts, and struts were made for me by pupils of the Engineering Institute (what better place could I have chosen!); and dope and paint for the wings were mixed up by the local chemist.

In the Dutch East Indies, where I landed in a field in which someone had planted long pointed stakes to mark out the design for a new house, the tears in the wing fabric that resulted from landing in their midst were patched up with pink sticking plaster.

Petrol-pumps, of course, had never been heard of in the out-of-the-way places where I wanted to refuel. The task of filling the tanks from huge casks was hard work indeed, mostly carried out in broiling heat, driving sand, or pouring rain.

And so I could go on; but I would not have you think that those early days were without their good side too. Everywhere I went an aeroplane was a novelty, something to be gazed at with awe and admiration. In India, for example, my unexpected landing was said to have prevented a native rebellion, as it was superstitiously believed the gods had intervened. Nowadays, I am afraid, record-breaking pilots are just a nuisance, something that disorganises the routine work of airports. Although conditions to-day are what we have all ostensibly been trying to achieve, yet I myself have to admit that I am sorry the bad old days are gone.

My flight was followed by countless others. Kingsford-Smith was the first to beat Hinkler's record. In October 1930 he left Heston, reaching Port Darwin 9 days 21 hours 40 minutes later. Many tried, but failed, to beat that record, but finally Charles Scott, in April 1931, cut 17 hours off 'Smithy's' time. The record was reduced gradually during the next three years, sometimes by a few hours, sometimes by a whole day, until in October 1934 Charles Scott and Tom Campbell Black in their De Havilland Comet smashed the record to pieces by their magnificent flight of 71 hours 18 seconds. This record stands to this day...

[AG Note: I decided to end this extract here. Amy goes on to outline how the big commercial airlines, such as KLM and Imperial, continued to expand their services to the Far East and Australia – the route she had flown herself nine years earlier.]

++++++++++++++++++

Chapter X: GLIMPSE INTO THE FUTURE (extract only)

It is good fun to guess 'What next?' To speculate on how fast and how high we shall fly to-morrow; to imagine a day when planes are as cheap and plentiful as motor-cars; to dream dreams of rocket flights to the planets for week-ends in a new world: and to wonder if the day will ever come when the aeroplane will be the tool of peace and progress and not a weapon of destruction.

Image 55: The abbreviation KLM means Koninklijke Luchtvaart Maatschappij – better known as Royal Dutch Airlines. Started in 1919, it is the oldest airline in the world. Courtesy KLM.

New and more advanced designs are germinating in the brains of plane and engine designers so soon as the old ones are pinned to the drawing-board. Scientists are busy edging back the limit-line of impossibility. One-time wonders are everyday commonplaces.

Rather than paint a Wellsian picture of 'Things to Come', I will try to indicate the more immediate future, something that is within our reach and who most of us will live to see, because, attractive as some of the forecasts are of speeds of a thousand miles an hour and round the world in twenty-four hours, I am afraid such visions are not backed up by present scientific knowledge, as I will try to show. Moreover, no one would care to prophesy anything really more definite than the fact that the next ten years of flying will far outstrip the last.

The Aeroplane in War.

Probably, in these times of a world's rush to rearm, the aeroplane takes first place in our minds as an instrument to kill. The finest brains, best material, and millions of money are being used to manufacture thousands of fast fighters, long-range bombers, quick-firing guns, high explosive torpedoes and bombs. Tens of thousands of men are being trained to outdo each other in dealing out death. Spain and China have served as the experimental aerial battlefields for future wars. We have grown used to reading of air-raids on helpless cities, of the slaughter of old men, women, and children, of fire and destruction, disease and pestilence.

Our warplanes are doing speeds of 362 m.p.h. (super Spitfires) and climbing to heights of 11,000 feet in five minutes, whilst our commercial planes are still hobbling along at a hundred miles an hour. The flower of our country's youth is being pressed into military service, whilst our commercial airlines and flying-schools are crying out for pilots and instructors.

New Royal Air Force flying-fields are being constructed daily, whilst our airports are being closed down one by one.

Where is it all going to end? In another war to end wars? Or in a war which will end us? Discussion and controversy rage backwards and forwards, like a shuttlecock across a net, on the relative virtues of the fighter versus the bomber; the utility or stupidity of the balloon barrage; the best methods of precaution against air-raids; the relative dangers of poison gases, incendiary bombs, and high explosive shells, and so on.

It is outside the scope of this book to go into the pros and cons of such arguments, nor have I the space or inclination to write of the horrors of poison gas, real and imagined, of death-rays, of fumes sprayed from the air to deprive the civilian population of movement and memory, of new death-dealing engines and metals, and such-like ghastly visions of the modern writer of fiction and plays.

Serious thinking men write reasoned truths. Professor J. B. S. Haldane, in a lecture he gave recently to the Royal United Services Institution. expressed the firm opinion that much nonsense is talked and written about the terrible things that are going to happen in future wars. He thinks it very unlikely that anything worse than mustard gas will be produced (some people, of course, wonder if there can be anything worse than mustard gas), or that explosives will get much worse, as there is a limit to the amount of energy which can be put into a given weight, or that disease will be spread by spraying microbes from aeroplanes. This is so difficult, he says, that high explosive shells will probably be found easier and more effective.

I have even read recently a case for the defence of bombing! The point of view taken is that, "If thirty thousand men, fine and young, straight and strong, die to-morrow in some battle it will rate but a paragraph in our news. But if a city is bombed and one thousand of the city's inhabitants are killed - the aged, the weak, the nerveless child - the headlines will scream.

"The war of the future is no longer confined to the battle zone, no longer can it be isolated to within a few miles of a given line. And with this fact slowly being discovered by all, I can see the ultimate arrival of a real and lasting peace. . The bombing aeroplane heralds a new peace" (T.Wewege-Smith, in The Aeroplane, 28 December 1938).

In any event, one fact, taught us by the wars of Abyssinia, Spain, and China, is indisputable, and that is, that the air arm, though it will undoubtedly play the most important part in any future conflict, cannot of itself decide the issue. Territory, to be won, must be occupied, and aeroplanes cannot occupy a territory. They might drop troops by parachute, but hardly in sufficient numbers to play a decisive part.

Let us hope that all this frantic planning and preparation, this pouring out of talent and money, may yet be turned to account for our country's welfare. In any event, for the purpose of looking into the future, I will take this point of view, for it is decidedly more pleasant to play the optimist.

Flying High.

First of all, let us take a glimpse upward into the stratosphere and guess how high we shall fly tomorrow.

Our present planes usually fly between heights of 2000 and 10,000 feet, because we have not yet in actual service the type of plane which can take advantage of those conditions which prevail at greater heights. Such machines can be built, and are in process of experimental construction, but nowhere yet is the high altitude transport aeroplane in actual use on an airline. It is, however, not very difficult to forecast that, within the next ten years, all the world's major air routes will be operated by such high-speed planes flying at great heights in regions of perpetually fine weather.

At the present time, the height at which a plane flies is usually dependent on the wind and the weather, and sometimes, too, on the regulations of the particular country over which it is flying. For example, in England, the minimum legal height recommended, though not enforced, that a single-engined plane should fly at a height of at least 6000 feet, so that in the event of engine failure he will have sufficient height to glide to the coast.

In America commercial planes have to fly at certain fixed altitudes. Flying westwards, they must be over 6000 feet altitude and keep on levels of even numbers of thousand feet. On the eastward trip, the minimum height required is 5000 feet, and planes must keep to the levels of odd numbers of thousand feet. This makes collisions on the busy trans-continental airways impossible. Usually American planes fly at greater heights than do ours, as designs are further advanced, and for many years variable pitch propellers, supercharged engines, and high octane fuel - all essential for altitude flying - have been in regular use. In fact, research work was started along these lines as long ago as 1920.

First, however, a word about the stratosphere, and in particular the substratosphere, where air routes of the future will mainly be found.

It is amazing how few people know that it is the balloon, and not the aeroplane, which has played the most important part in the conquest of the stratosphere. This can readily be understood once it is realised that the aeroplane depends on air for its propulsion, engine, and cooling, and to give it 'lift' to fly at all. As everyone knows, the higher you go the less air there is, and therefore an aeroplane's 'ceiling' necessarily depends on the amount of air it needs and the amount it can get.

The limiting factor of the ceiling of a balloon is totally different. It depends principally on the size of the balloon. The weight is of importance, too. because, although the lifting power of the bag always stays in front of its weight (on the theory that balloon volume increases by the cube of the diameter, whilst the weight increases only by the square), yet it is useless for a balloon to be sent up without scientific instruments, ballast (without which the balloon cannot descend safely), a crew. and a gondola. It is not strictly necessary for men to go up in the balloon, but there will always be scientists, like Professor Piccard, who must explore in person, and in any event they can put their experiences into words to humanise the figures and data of instruments. In addition to this, a 'manned' balloon can stay aloft longer, and from the scientific point of view this is important for the study of cosmic rays.

Small unmanned balloons are sent up, equipped with 'robot' observers, scientific instruments, and mechanical and photographic recording apparatus. The balloon stays up for three to four hours, and, when it reaches a certain height, it bursts and the instruments glide down to the ground attached to parachutes.

On 19th November 1938 six such balloons were sent up from the Franklin Institute, Philadelphia, U.S.A. It was ingeniously contrived that the instruments carried should communicate their readings to two radio ground stations by means of electrical impulses to a tiny radio transmitter. One of the balloons was actually recovered on 21st November in the Atlantic Ocean, eight miles off the coast of Massachusetts, with a note in a waterproof case asking for it to be returned to the Institute.

A balloon goes up only because it weighs less than the air it displaces. Therefore, not only is its size controlled by the weight it has to carry, as explained above, but a limit to this size is quickly reached as it climbs into the upper reaches of thin air, where gigantic dimensions would be needed.

Till now, rubberised cotton has been used for the fabric of the stratosphere balloon on account of its strength. If some lighter, but equally strong, material could be found, the weight of the bag would be reduced considerably and the altitude record would conceivably creep up a little. Expert opinion, however, is that the present record of approximately 14 miles will not be exceeded by more than a mile at most. It is calculated that to double the present record 2500 tons weight would have to be lifted.

Professor Piccard, however, is more optimistic and has a still unsatisfied ambition to reach a height of 30,000 metres (18.641 miles).

The name of Professor Auguste Piccard, a Swiss Professor at the University of Brussels, is famous the world over for his daring ascents into the stratosphere. It was his first ascent, in August 1931, with his assistant Paul Kipfer, which really started a new era for the balloon - that of stratosphere research. He rose to a height of 9.81 miles - the first man to enter the stratosphere.

Previous attempts had failed principally because the crews had perished from lack of oxygen. Professor Piccard solved the problem with a sealed gondola, in which the air pressure was kept at normal level, just as it is in a submarine, except that for deep-sea work the walls of the submarine must be made to withstand terrific pressure from without, whilst the stratosphere machine, or gondola, has to be strong enough to keep from bursting from the greater pressure inside.

The primary purpose of the balloonist is scientific, but the aeroplane seeks the higher levels for military and economic purposes. The balloonist, therefore, besides obtaining information on all kinds of rays which vitally influence not only mankind, but our climate and vegetable life, is helping aeroplane and engine designers by providing them with valuable observations on air pressures, densities, temperatures, and the like, as well as a knowledge of meteorological conditions in the upper regions which are important to the stratosphere navigator.

The Rocket.

As I have tried to explain, there is a limit to both aeroplane and balloon exploration, and, when that is reached, the rocket will come into its own and carry on an apparently limitless investigation. Whether we shall ever have week-end flights to the planets it is impossible to forecast, but the limiting factor would appear to be not so much the difficulty of getting there as the impossibility of living there once you had arrived. It would seem fairly certain, however, that rocket flights from place to place on our own globe are within the realms of possibility. Every day I seem to be picking up books in which something or other which has been deemed 'unlikely' or 'impossible' has nevertheless come to pass. Very often certain forecasts of ancient times, rejected and ridiculed at the time, are the very ones to be translated into truth to-day. It would seem stupid, therefore, either to repudiate or to scoff at some of the seemingly fantastic visions of to-day.

For example, H. G. Wells in The First Men in the Moon causes gravity to be overcome so that a rocket flight can be made to the moon. An engineer discovers a substance which renders weightless all objects which come under its shadow. That is all very well for a novel, but scientists say that it is impossible to remove gravity, and that to break through the earth's gravitational field an initial velocity of 12,250 yards per second is needed, an acceleration which no human being could endure. And yet - experiments are being made even to-day to overcome gravity, that strange force which we do not understand but whose power we decidedly feel and which so successfully limits our gropings into the infinite. From wishes and dreams often materialises an idea which, as science progresses, finally becomes a crystallised fact, as in very truth happened in the conquest of the air itself.

Rockets figure largely in myths and legends of the past. Sir Isaac Newton then came to wave the wand of science over this vision of conquering interstellar space. To-day his successors translate his scientific theory into the action of the rocket. Experiments are actually being carried out and rocket flights have been made in many countries, notably by Professor Goddard in America, but the 'Great Flight to the Moon' still remains unaccomplished.

Professor Piccard believes that the splitting of the atom and the harnessing of its powers to science will be the means of supplying the key to the problem of interplanetary flight.

With a rocket-ship a modern Jules Verne could circle the globe in twenty-four hours, cross the Atlantic in three, breakfast in England and dine in Australia.

The 'Stratosphere' Plane.

Before giving a few details of the experimental passenger-carrying 'stratosphere' planes being built to-day, just let me say a few words about the stratosphere itself, so that you can appreciate some of the problems involved.

The air is warmest near the earth, due to contact with its surface, the air transmitting heat at a greater rate than it absorbs it, and therefore the further away the air is from the earth's surface the cooler it becomes. The temperature falls on an average at the rate of 3 degrees per 1000 feet of altitude....

[AG Note: I decided to omit much of this Stratosphere section. Amy goes into technical, specialist detail about the extent of the stratosphere, the height planes could and could not fly, cabin pressure, speed, and how Britain was lagging behind the USA in night-flights, etc.]

++++++ +++++++++++++

Britain is second to none so far as her flying-boats are concerned, and the famous company of Short Brothers is building huge flying-boats for Imperial Airways service across the North Atlantic. These boats are being designed to refuel in the air. As a matter of interest, so keen is the competition to be amongst the passengers on the first commercial crossing, that Imperial Airways have already a hundred people down on the waiting-list. Such is the faith we have to-day in aviation!

So far as the passengers themselves are concerned, high altitude flying will be far more comfortable for them, as they will be above most of the 'bumps' which make flying in the lower atmosphere sometimes so unpleasant. As against this, the monotony and boredom of a long trans-ocean flight will be great, and it is certain that some forms of amusement will have to be contrived to while away the tedious hours.

Image 56: Imperial Airways Flying-boat. Imperial Airways operated from 1924 to 1939, served parts of Europe, but especially British Empire routes to South Africa, India and the Far East to Malaya and Hong Kong. They merged into the British Overseas Airways Corporation (BOAC) in 1939. Flying boat always impressed me when I was a boy; so I was disappointed when they faded off the scene. Courtesy Wikipedia.

Although, theoretically, there is no limit to the size of an aircraft, yet it is the general opinion amongst aircraft designers and airline operators that a larger number of smaller planes making a high frequency of service possible is more economical and practical than a few huge unwieldy machines costing a fortune to build and maintain. (It should be noticed that flying-boats need not be limited in size by consideration of aerodrome problems, as landplanes necessarily are, which is one of the reasons why flying-boats of all countries are, in general, larger than their landplanes.)

Because of this limiting factor of aerodromes it is likely that landplanes will not exceed some 50 to 100 tons for the next twenty-five years, whilst flying-boats are already being planned to carry 100 tons and more.

Such a boat could cross the Atlantic in less than twenty hours, and would have comfortable staterooms, a dining saloon where games and dancing would take place in the evenings, promenade decks, smoking lounges, a library, and, in general, such luxuries as you would normally expect to find on a yacht or first-class liner.

To sum up, therefore, it would seem that the tendency of the immediate future is to increase the size, speed, 'ceiling', and range of commercial aircraft, all such factors, however, being more limited in landplane than in flying-boat construction. The landplane design must still be a matter of compromise between airport and aircraft designers, whilst flying-boats will only be limited by questions of economy and practicability. However, it is always dangerous to forecast in these days of rapid change, and the discovery of some new substance or metal or the use of some new fuel, such as liquid hydrogen, would revolutionise the whole picture.

Future Air Routes.

It is easy enough to work out in theory how quickly we can get from place to place, given the speed of the plane and the distance between the two places. It is also easy to draw on the map a straight line between these two places and call this the air route.

In practice, however, the average speed works out considerably lower, and the straight line often has to be given a decided kink.

From the actual cruising speed of the plane must be deducted time for 'delays,' such as spending the night in hotels instead of in a sleeping berth while being speeded on your way (this is where we lose the most time on our own trunk airlines, though we could avoid it if the routes were equipped for night-flying); refuelling (essential delay, but could be speeded up considerably); making detours to pick up passengers or to land on a suitable aerodrome for refuelling (these 'delays' could be avoided by ' feeder' lines and more and better aerodromes); time wasted in climbing to cross some mountain range or to clear a patch of bad weather (could be avoided if high altitude machines were used, when not only would their operation be the most economical but at the same time all obstacles of earth and weather would be safely cleared).

K.L.M. are now planning a twice-weekly service between Croydon and Australia, to commence early in 1940, with a service of 3½ days in each direction. The present time is eight days. The latest Douglas landplanes will be used, with accommodation for 42 passengers by day and 24 by night in sleeping berths. All-day and all-night flying will, of course, be the rule....

[AG Note: Once again, I decided to cut out technical details – too much for my humble brain].

++++++++++++++++++

The actual cost of learning to fly has recently been considerably reduced by the inauguration of the Government's Civil Air Guard scheme, but even this is a scheme designed rather to provide a large reserve of pilots for military purposes than merely to foster private flying, nor does it in any way help the newly-fledged pilot to own his machine or even to hire one apart from his ordinary school training.

Therefore, the cost of learning to fly still remains prohibitively high for the majority wishful of flying for business or merely for pleasure, and the cost of buying and of maintaining a private aeroplane is impossible for all but a very wealthy few. I have owned planes myself and I know what I am talking about. Not only is there the initial cost of the plane, but in addition there are maintenance charges (compulsory, especially the yearly overhaul for the Certificate of Airworthiness), cost of high-grade petrol (much of which is tax, ostensibly for the Road Fund! In France, this is returned to the pilot) and oil, insurance (extremely high), hangarage fees (far higher than garage charges), landing fees, and all the other expenses coincident with travelling.

Added to all this expense there are such problems as finding your way around. Private users of wireless are not welcomed and in any event have to take second place if an airliner needs attention, and, whilst aerial maps are in every way excellent, yet finding your way in poor visibility is very hazardous, as witness the number of forced landings and crashes due to people getting lost and coming down low to read the name on some railway station or making an actual landing in some field to make inquiries. Aerodromes are too scarce and the time taken to get to your ultimate destination even after you have safely landed is too long. These and other troubles could quite easily be solved with a little official help and co-operation. For example, it would be very easy to sign-post the country by painting the names of towns on railway sleepers (easily blacked out in time of emergency).

Personally, I do not now own an aeroplane because it is too expensive. I glide, however, because it is peaceful, cheap and much more interesting than 'power' flying.

Apart from the contentious question of private flying, there is no doubt whatever that aviation is a major force in our national, economic and military life, and that if we do not recognise this we are not in step with the times. My most fervent wish is that the aeroplane will very soon have its chance to develop as an instrument to foster international trade and to enrich a lasting Peace.

~~~~~~~~~~~~~~~~

AG Comments:

I was hoping to draw on much more autobiographical information from this publication, but there was not as much as I expected. Primarily, Amy's Sky Roads book focused on: how the airline routes were first opened up; what airports and airlines were like in the 1920s/30s; and speculated about the great potential for the future of civilian aircraft flying around the world.

What Amy could not be expected to predict, however, was the full impact of the pending Second World War (she was writing in 1938/39) upon aircraft development, nor the vast advances provided by jet propulsion engines, helicopters, guided missiles, or the atomic bomb.

Amy's Dedication in 1939 sounds, with the benefit of hindsight, to be partly prophetic when applied to herself. Sadly, within a couple of years, she too had fallen 'by the airwayside' – and writers have ever since been challenged about how she died.

Her Preface has a quaint and marvellous disclaimer against any aviation experts criticising her writing. She had the celebrity status and privilege to be able to state the book is meant to be "rather a romantic story of the world's great sky routes" (p7). She is right, of course, to focus on the "human drama" which pioneers like herself endured to open up the sky roads and enable the big multi-nationals to follow in her vapour trail. The major drawback from Amy's perspective, however, was that they drained away the "magic of the air" which was there in the 1920s and early 30s.

I include the full list of Amy's ten chapter titles to whet any reader's appetite should they be tempted to read the whole of Amy's book. I selected extracts from only three of her chapters: I. The Dawn of the Air Age; II. To Australia and New Zealand; and X. Glimpse into the Future.

In her first 'Dawn' chapter, she was certainly a big fan of the Wright Brothers. That must have been a dramatic meeting between Amy and Orville Wright (c.1935) – she was the star-struck superstar. Actually, I for one did not realise that Britain purchased the Wrights' patent on flying for £15,000. How did that work? Did the UK Treasury earn royalties every time a plane took to the air? I must visit the London Science Museum to see the Wrights' plane on display. I might even go and see Jason – assuming it has not been given to Kingston-upon-Hull on a permanent basis (we can all dream!).

I included a large extract of Amy's second chapter where she focusses on her flight to Australia in May 1930. I was delighted to read about her attraction for undertaking such a venture. Her romantic side added to her motivation: the glamour of hair-raising stories, torture-loving bandits, wild beast, cannibals, and sharks – they all fired her "school-girl imagination" (p41). Esoteric aside: was Amy's romantic and childhood attraction to "torture-loving bandits" linked to her being drawn to men who treated her badly?

Along with her happy memories of Arabian Nights, she mentioned Greek Mythology. I would have been over the moon had she also written about Classical figures likes Ceres (H.219), Achilles (H.109), Hermes (H.209) or Melpomene (H.1474) as part of her grandfather's trawler fleet bearing Classical Greek names. She must have known her family once owned fishing boats.

On the other hand, she openly acknowledged the splendid "invaluable financial help" for her Australia flight from Lord Wakefield of Castrol. Yet omitted to make one mention of Jason Kippers or Andrew Johnson Knudtzon - whose profits purchased her Gipsy Moth in the first place. Why?

Image 57: Lord Wakefield (1859-1941) owned Castrol Oil. The name Castrol arose because castor oil was added to the company's lubricating oils. Coincidentally, philanthropist Charles Cheers Wakefield died ten days after Amy was lost. Courtesy Castrol Oil.

Amy was keen to contrast the ups and downs of her 1930 venture with the comforts enjoyed by modern-day passengers. For example, the problems she experienced with maps, equipment, permits, petrol, weather reports (or lack of them) and her various crashes along the way – all of which she overcame magnificently.

Nostalgically, Amy's biggest regret was that "record-breaking pilots are just a nuisance…I am sorry the bad old days are gone" (p50) – though she might have been tongue-in-cheek when she wrote this.

In her final chapter (Glimpse into the Future), Amy displayed her incredible ability to be free- and far-thinking as she speculated about where aviation was progressing. To her credit, she even explored the possibilities of deep-space travel. Impressively, her section about The Rocket, concluded: "the splitting of the atom and the harnessing of its powers to science will be the means of supplying the key to the problem of interplanetary flight" (p295).

She was above any blind, patriotic jingoism of her age in that she prudently saw the terror of war too. In 1939, Amy was still one of the few female pilots. It was good to see her express horror at the menacing war that was on the horizon. She accurately assessed that the warmongers merely used Abyssinia, Spain and China as "the experimental aerial battlefields for future wars" (p285). That, indeed, proved very true and demonstrated how perceptive and prophetic she was in her analysis. Her strong sympathy was with the women and children in the "helpless cities".

Added to this, she predicted what military experts have subsequently concluded about the aerial bombing of Vietnam and Syria: that air power alone "cannot of itself decide the issue. Territory...must be occupied, and aeroplanes cannot occupy territory" (p288). In other words, 'boots on the ground' – as we say today. I have often praised Amy's sense of humour, but I am also delighted to see her anger in print – a wise woman.

She certainly embraced complex scientific issues when, for example, she seriously considered the stratosphere in great depth. As well as being a qualified engineer, I believe Amy could easily have become a scientist – had she lived long enough and given the right opportunities (perhaps at a campus in the States).

It is not surprising that Amy goes into technical detail – after all, she was an engineer of note and her mind was naturally drawn to the mechanical side of any issue connected with flight. Had Amy survived the war, I feel she would have rejoiced in the aeronautical advances of supersonic flight. Furthermore, her anti-war sentiments might have led her to join the CND (Campaign for Nuclear Disarmament) movement and dabbled on the fringes of politics. Who knows? As a former President of the Women's Engineering Society, she would have become a venerated figure on aviation matters, been enlisted onto Government panels, and continued as a strong advocate for women in all fields of engineering.

One of my motivations for including these Appendices is primarily to encourage readers to read Amy at source – and take their understanding of her life much further. Amy ended her one and only full-length book on a personal and positive note. She admitted that she could not afford to run an aircraft of her own. Instead, she was happy with her newly found love of gliding. At her usual optimistic level (though immediate events proved her hopelessly wrong), she concluded "My most fervent wish is that the aeroplane...[will] foster international trade and enrich a lasting Peace" (p314).

################

(15E) A Day's Work at the ATA by Amy Johnson

Appendix 15E: JOHNSON, Amy. 1 January 1941. Chapter + Letter. A Day's Work in the ATA. Woman Engineer: the Organ of the Women's Engineering Society. March 1941. Journal. Amy Johnson – In Memoriam. 5. 6. 89-90.

The shrill note of the alarm blares in the sleepy ear of Miss X, ferry pilot of the Air Transport Auxiliary [ATA]. Confused by sirens and "all-clears," she doesn't realise for the moment that this new noise is merely the signal to get up. At last the alarm's insistent call penetrates her brain and she knows the day's work has begun.

AETHERIS AVIDI

Image 58: The Latin motto of the Air Transport Auxiliary Aetheris Avidi means "Eager for the air". Courtesy ATA Association.

The clock points to 7. Outside it is pitch dark and it is impossible to see what the weather is likely to be. If there is fog there is no hurry, and she would dare roll over and have that extra, fifteen minutes. Unfortunately, however, it is much too dark to see, so she has to make the best of things and get up quickly in case it is fine, as there is certain to be lots of work to do.

In the winter months, aircraft production and repair work go on unabated, increasing every day, but the bad weather holds up delivery of machines from factories to aerodromes, and numbers of planes accumulate, making no mean problem for the organisation staff, as, somehow or other, in spite of the weather, the machines must be dispersed in order not to present too alluring a target to the enemy.

Miss X hastily puts on her navy uniform — slacks, military tunic, forage cap and overcoat, pausing just long enough before the mirror to admire her two new gold stripes. She is now a First Officer, and is qualified to fly a great many more types of planes than — alas! — her employers BOAC (British Overseas Airways Corporation – under whose aegis the ATA has its being) will allow.

"All single-engine Service types and light multi-engine trainers" is the caption on her C.F.S. [Central Flying School] report. However, no use wasting time sighing over the Hurricanes and Spitfires she is not permitted to handle — rather is she grateful for the enormous privileges she has so far gained. The war has placed opportunity right in her path. At no expense to herself, she has been given the best training that money can buy — an R.A.F. Central Flying School Conversion Course. From the light training machines, like Moth and Avian, she flew before the war at her local Flying Club, she has been 'converted' to fast modern types with their hundreds of knobs and complications. Thanking her lucky stars, she eats a hearty breakfast, as the chances are she may not have time for any lunch. The aerodrome is a sharp ten minutes' walk away from her digs, and she steps out into a misty wet fog which tries its best to dampen her high spirits.

Arrived at the aerodrome, her first enquiry from the Duty Pilot (who has, poor soul, been on duty since 7.30 a.m.) is what are the chances of the weather clearing. "In an hour's time" she is told optimistically. The next enquiry is the vitally important question — What exciting sort of aeroplane has she been allotted for her first day as a First Officer? "A Tiger Moth from Y to Z" she is told. Stifling a bitter disappointment, she goes to the locker room to prepare her maps; enquires from the Signals Officer (Miss Susan Slade, late secretary of Airwork, Ltd., Heston, or Miss Connie Leathart, veteran peace-time private owner of wide experience) where are the balloon danger zones on her route; obtains from the safe the "Signals" for the day (i.e., signals which must be given if one's plane is challenged by one of our own aircraft, ships or ground stations); telephones for a route weather forecast; asks permission from Fighter Command H.Q. to land at the aerodrome for which she is bound; collects together her parachute, Sidcot suit, flying boots, gloves, helmet and goggles, emergency kit for the night (in case she cannot get back to her base), some sandwiches in lieu of lunch, and, reinforced by a cup of hot coffee from the canteen, is ready to start.

Unfortunately, the weather still persists in being damp and misty, and only an enthusiastic pupil of the Women's Section's instructor — Miss Margaret Cunnison, formerly chief instructor at Perth — takes off to do a "circuit and bump". Miss X, tired of gazing at the miserable sky, goes inside into the cheery atmosphere of the "mess" and challenges someone to a game of darts. The game finishes, two others are played, more coffee is consumed, knitting is taken out, friendly gossip and "shop" fly back and forth, one or two of the girls ask for and receive permission to go and have their hair washed and do some shopping.

"How many hours did you get in last month?" asks Miss X of "Margy" Fairweather, one of the four taxi pilots. "70 hours on Ansons on taxi work and three hours on a ferry flight to Scotland." "You lucky thing" chimes in Mona Friedlander, erstwhile ice-hockey champion. "I only manage to get in about thirty hours' ferrying, with at least a hundred waiting about!" Margy smiled quietly to herself, as there are two definite points of view on this subject. An outsider would naturally conclude that the more hours flying a pilot puts in in a given time the harder she must undoubtedly have worked, but the outsider "knows nuthin". Amongst the ferry pilots (i.e., those delivering machines) competition is as keen as mustard for any jobs that may be going, and the hours flown are carefully collected in a most miserly fashion. The smaller their total the keener they are to augment them, only those with two thousand or more losing to some degree this urge to "pile them up". The taxi pilots, on the contrary, always have so much extra flying that their totals for the month are always in excess of anyone else's. For example, a ferry pilot may have a job which means only twenty minutes' flying for the day, whilst the rest of her time is spent waiting on the aerodrome for the taxi machine to collect her, or riding with the taxi pilot on her long round. The outsider would consider twenty minutes' flying a very slack day's work as compared with the taxi pilot's probable five hours (or even eight or nine in the summer), but ask both the pilots separately what they think! Anyway, Margy Fairweather won't swop her taxi job for even the most exciting machines the ferry pilots fly, which is a very good thing for the rest of the girls, as she is one of the safest and best pilots to be found anywhere in the country, and taxi work is a great responsibility, to say nothing of being in the nature of a command performance before an audience of experts.

Miss X, however, is getting tired of talk, and wanders outside to have another look at the weather. It looks rather better, so she asks the Chief Ferry Officer, if she can "have a crack at it". The C.F.O. confers with her C.O., Miss Pauline Gower, well known as to her high flying capabilities but nowadays seen mostly in the light of a clever psychologist, studying and understanding to no mean degree the temperaments of the girls under her care. The C.O. knows that Miss X is a good, careful pilot, not brilliant, but having the common sense to know her limitations, and she feels justified in allowing her to set off if she feels like it, knowing that she won't take undue risks. After all, the work of the A.T.A. is to deliver machines safely and in one piece. Whether it be to-day or to-morrow matters far less than the condition in which a brand-new highly expensive machine arrives.

All Miss X has to do now is to persuade the taxi pilot to take her to the aerodrome from which she has to collect her machine. Her pilot being Mrs. Fairweather, this is easy. Arrived at their destination, Miss X gets out, with parachute, Sidcot suit, maps and kit and goes in search of the Station Engineer. Eventually, after innumerable papers are signed, she is tucked into the cockpit of a shining new Moth, the propeller is swung and she is away.

Even if she does have to battle in an open cockpit with wind and rain, snow and hail; though she may lose her way in driving mist and narrowly miss colliding with a balloon barrage; though she may at last arrive frozen and frightened, she knows it is useless to "shoot a line" to people who are doing this sort of thing every day as a matter of course. So she just gets her receipt signed and makes enquiries about transport back to her base. If she is lucky, the taxi machine will come and collect her, though there may be a long wait, but otherwise she will have to "hitch-hike" or take train, 'bus and car. Only too often a half-hour flight entails hours of travel to get back to her base, but she is as used to this as she is to the uselessness of airing her troubles.

Back at last, she triumphantly hands over her precious receipt for the safe delivery of her machine, and, after packing away her kit in the locker room, putting her parachute on its proper place on the shelf, and locking away her maps, she finally goes home, to what she is herself at any rate convinced is a well-earned dinner and sleep (perhaps to dream of the super machine she may have to-morrow).

Back at the aerodrome, in the Operations Room, her day's work is officially entered up as "One Tiger Moth, Number . . . . delivered by First Officer X (followed by appropriate times and places). Flying Time fifty minutes". Just another job done.

~~~~~~~~~~~~~~~~~

AG Comments:

This is an interesting account of the daily routine of a typical ATA pilot. I do not know if many records exist by female aviators in WWII; but this certainly adds to the social history of wartime Britain. Written in third person (Miss X), this is a good account by Amy of what she did during a typical day in war service. As well as the excitement of flying and the satisfaction of a job well done, she also described the boredom of sitting around waiting for the weather to improve; the small talk; playing darts; knitting; and, especially the camaraderie amongst the airwomen.

There is a poignant moment in this piece where Amy mentioned that an ATA pilot who, in the wind and rain, the snow and hail, "may lose her way in the driving mist" – just as she did a few days later in that fateful January 1941.

Amy pointed out that once an aircraft had been delivered, the poor pilot then had to find her own way back to base if there was no "taxi machine" going her way. Given those hard-luck circumstances, she might take a train, bus, car or even hitch-hike. This evokes quite an image in my mind: the world-famous Amy Johnson thumbing a lift in the middle of the war.

Despite her high social standing, Amy displayed a "no fuss, just get-the-job-done attitude". Miss X is down to earth, pragmatic, and keeps at it until the job is done. To my mind, this echoes a typical attitude / approach on the Hull Fish Dock. If there was a heavy landing of fish, there was no procrastination, the job had to be done and no one went home until it was all finished. That fish dock attitude was clearly expressed by none other than the legendary Hull F.C. rugby star and Hessle Roader Johnnie Whiteley. His working life began as a barrow lad with Fields Fish Merchants and he reminisced that "One of the lovely things that transpires from working on the fish dock was you worked flat out or belly-to-ground...you never knew anything else but to work and get finished" (Gill. 2010a. 15:30-16:16). It seems to me that Amy also absorbed that sort of work ethic from being involved (indirectly) with the family fish trade.

#################

(15F) Letter to Women's Engineering Society by Amy Johnson

Appendix 15F: JOHNSON, Amy. 1 January 1941. Letter. A Day's Work in the ATA. Woman Engineer: the Organ of the Women's Engineering Society. March 1941. Journal. Amy Johnson – In Memoriam. 5. 6. 88.

1st January, 1941.

My dear Caroline,

Enclosed is my effort for the "Woman Engineer" which you asked me to write. It is an honour to be asked to contribute to your magazine, and I am afraid my article isn't anywhere near adequate. My only excuse is that I am so out of practice (this is my first bit of writing since war started) that I find it extremely difficult to string words together. I'd rather do a week's flying any time!

Very many thanks for your card for Xmas. I'm afraid I didn't send any out this year, but I certainly appreciated being remembered by my friends.

I hope the gods will watch over you this year, and I wish you all the best of luck (the only useful thing not yet taxed!)

All the best,
Yours affectionately,

NOTE: This last letter from Miss Johnson, written a few days before her death, is so typical of her that it seems justifiable to share it with her friends in the [Women's Engineering] Society. Caroline Haslett, President

Image 59: This is the cover image of The Woman Engineer journal for March 1941, Volume 5, Number 6. The Society formed in 1919. One of their primary aims is to support women to achieve their potential as engineers, applied scientists and leaders and to reward excellence. Courtesy Women's Engineering Society.

AG Comments:

Amy certainly displayed a modest, deferential stance to the editor of the Woman Engineer Journal. She is semi-apologetic about her piece. Interestingly, for me at least, is her allusion to "the gods" (rather than any monotheistic god). I am not for one moment suggesting that she was a committed pagan, but it does fit into my (recurring) theme of seeing Amy as being highly superstitious in her outlook on life. Those who adopt folklore beliefs unwittingly acknowledge multiple gods of this, that, and the other when 'touching wood' or invoking the sea gods for protection. In pre-Christian times, magical allusions were made to the stars, moon, tree spirits, Jove or whatever. Similarly, Amy also uses the modern-day term of Xmas rather than the respectful Christmas. Regardless of that, I love her sense of humour about the government not yet being able to tax "LUCK".

##################

(15G) Obituary Letter to The Times by Caroline Haslett

Appendix 15G: HASLETT, Caroline. 14 January 1941. Obituary. Letter to the Times. Reproduced in Woman Engineer: the Organ of the Women's Engineering Society. March 1941. Journal. Amy Johnson – In Memoriam. 5. 6. 84.

All the world knows of the Amy Johnson who flew solo to Australia ten years ago, but it is perhaps those who knew her more closely who were able to appreciate her gifts and abilities, the generosity of her mind, her modesty over real achievement, her unquenchable spirit which, with her keen wit and boundless humour, must have carried her through times of tedium as well as of horrific experience. Whatever Amy did she did it with zest and relish. The sparkle and vigour of her personality communicated itself to all who came into contact with her, and the Women's Engineering Society enjoyed it in full measure during her three years as its president. She was no nominal president, but someone who imparted her own verve and enterprise to this society, to whose pioneering spirit her own was akin. She was always ready to give of her time and talent, and the latter certainly was not limited. As a public speaker and as a writer she had a clear, incisive style, and the ability to infect others with her own enthusiasm.

Amy Johnson was intensely alive to the beauty and strangeness of form and colour which her flying experience presented to her in a very vivid manner. Her book, Sky Roads of the World, is full of many word pictures seen from the cockpit of her aeroplane, and she infused into them the emotions she must have felt when enchanted by the vagaries of sea, sky, and cloud, or awe-struck by the cruel and relentless manifestations of nature in adverse mood. The élan which characterised Amy's activities either in word or deed was tempered by a shrewd common sense; the vision which inspired and the ardour which led her to attempt her feats of aviation were accompanied by a capacity for endurance which is not always appreciated by those who read of the triumphal conclusion to a well-nigh impossible venture. Only recently the Woman Engineer received from her an article telling of the pleasure she found in her work in the Air Transport Auxiliary and of her delight at the opportunities it gave her, an article which, while recording the satisfaction in a job well done, exults in the unexpected turns encountered in the performance, and in the camaraderie to be met on every hand — its author was truly Amy.

CAROLINE HASLETT
The Times, January 14th, 1941.

~~~~~~~~~~~~~~~~

AG Comments:

Caroline Haslett summed up a wonderful batch of Amy's traits: generosity, modesty, unquenchable spirit, keen wit, boundless humour, zest, sparkling personality, enterprise, enthusiasm, shrewd common sense, and a capacity for endurance.

I believe this positive list were not merely a sentimental reaction to the loss of a close friend, but a genuine account by an astute, intelligent society president. Equally, given my angle, I would like to believe that Amy acquired many of her playful qualities during her Hessle Road tomboy days.

It was also good to see the President of the W.E.S. applaud Sky Roads of the World. She highlighted Amy's "word pictures seen from the cockpit ...enchanted by the vagaries of sea, sky, and cloud, or awe-struck by the cruel and relentless manifestations of nature in adverse mood". Praise, indeed, for Amy the author.

##################

(15H) Obituary Letter to The Times by Pauline Gower

Appendix 15H: GOWER, Pauline. 8 January 1941. Obituary. Letter to the Times. Reproduced in Woman Engineer: the Organ of the Women's Engineering Society. March 1941. Journal. Amy Johnson – In Memoriam. 5. 6. 84.

Miss Amy Johnson was known throughout the world for her many famous flights, but in her private life and as a person she was less well known. I had the privilege of meeting her first in 1930, and during the years that followed got to know her as a friend. Although I always appreciated her brilliance as a pilot, the attributes which went to make her character were to me more impressive than her wonderful feats in the world of aeronautics. Her physical courage as an aviator was – undoubted; her moral courage, her large-heartedness and her sense of humour were only fully appreciated by her friends. In her private life Amy Johnson was unassuming and entirely lacking in conceit. It is inevitable that the name of someone as famous as she was should be coupled with many extravagant stories. Those who knew Amy Johnson intimately saw her as an ordinary human being, keen on her job, brilliantly successful but always accessible. After her spectacular flight, when the world was at her feet, she could spare the time to give encouragement, help, and advice to any who asked her. Many have been assisted, encouraged, and cheered by her.

When she joined the Air Transport Auxiliary she settled down to her new life with all the eagerness and enthusiasm of somebody who obviously had her heart in her work and was anxious to do a job for her country. The flying she was required to do was not spectacular, but it required steady application. Sometimes it was easy for one with her experience; at other times her skill stood her in good stead. Whatever the circumstances, however she was feeling, the job was done; and the conscientious manner in which she carried out her duties was an inspiration to all those who worked with her. Amy Johnson is not only a loss to aviation; those who knew her have lost the type of friend who cannot be replaced.

PAULINE GOWER.
The Times, January 8th, 1941.

~~~~~~~~~~~~~~~~

AG Comments:

Pauline Gower makes a good distinction between the public and private faces of Amy. Pauline tapped into Amy's deeper characteristics: physical and moral courage, unassuming, keen to get the job done, patriotic, steady application and (yet again) her sense of humour. Many of these positive traits, especially her sense of fun, were clearly manifested during Amy's youthful, pre-university days.

##################

(15I) Anonymous Obituary by an ATA Pilot

Appendix 15I: Anonymous. March 1941. Obituary. Amy by a fellow pilot of the ATA. Woman Engineer: the Organ of the Women's Engineering Society. March 1941. Journal. Amy Johnson – In Memoriam. 5. 6. 85.

The two most noticeable things about Amy Johnson were, at one end of the scale, the range, depth and consistency of the devotion she inspired in many thousands of people who could never ordinarily hope to meet her, and, at the other end, the affection of those who knew her well. She had her detractors, but they fell mainly between these two classes and consisted mostly of the smaller fry of aviation.

To fly regularly with Amy was a revelation; ten years after she had made the flight which, in fact, made her, she had only to land at an R.A.F. aerodrome for airmen to crowd round. Some of them only wanted to look at her, but many wanted her autograph. In fact, it became a kind of Air Transport Auxiliary "family" joke that whenever she was to be in the party, an extra ten minutes must be allowed at every stopping place for Amy to sign autographs. The officers would want to take her out to lunch — a superior kind of autograph hunting.

One day, at a country inn, a party of the A.T.A. were having lunch. The cook came in, carrying a six months' old infant: would Miss Johnson hold the baby, just for a minute? A week after her death, two women ferry pilots were dining at a well-known restaurant; the Chef brought to them personally a new savoury he had created, "in memory of Amy Johnson". Many such stories could be told by her colleagues, but perhaps the finest tribute of all was the unconscious one paid by a member of the congregation at her Memorial Service. Half a dozen women A.T.A. pilots, in uniform, were acting as ushers; someone was heard to ask who and what they were. Amy was bigger than the thing to which she belonged.

The secret of Amy's popularity lay, I think, in a combination of two facts. Firstly, her achievements were not made easy by advantages of wealth, position or influence. She got away, and got away by her own efforts, from the treadmill which makes up the life of the ordinary young man and young woman, whose escape can usually be only a vicarious one through the celluloid adventures of the cinema. Amy typified Hollywood come true. The second fact was the general feeling (it is surprising how accurate most popular estimates of character in public figures are) that she was not out for what she could get. It was somehow sensed that achievement was itself her objective, and that although she did reap big rewards, and certainly enjoyed getting them, they were quite incidental. It was for these two reasons, taken together, that she captured the imagination of the world.

Timing had of course much to do with her fame; a few years earlier would have been too soon, and a few years afterwards, too late. That this had little to do with her popularity, as apart from fame, is proved by the comparative oblivion into which most of the other record-breaking pilots have sunk.

Many people have wondered why Amy did not receive the public appointments which have fallen to other women in aviation. I think the answer lies quite simply in one of the things which endeared her to so many people; she had no eye to the main chance. If you want a thing enough (and the operative word is "enough") you know instinctively how to conduct yourself so that it comes your way. Undoubtedly, too, she did some silly things at times — the sort of things which office-seekers simply must not do — but then why the devil shouldn't she? The things which Amy did want were friends, flying and fun, and these she had in full measure.

She gave much, too. As a senior pilot of the Women's Section of the A.T.A., she was kindly and helpful to the younger ones, generous about the achievements of her equals, and completely silent about her own. She was so direct and unassuming that they all quickly forgot she was a celebrity, and learnt to accept her as one of themselves. Perhaps the fact that for probably the first time she was working with pilots who could have no sex jealousy may have helped.

Much of her time was spent in giving conscientious replies to the hundreds, mostly women, who wrote to her for advice about flying careers; and she tried in many ways to advance the standing of women in professions generally, and in aviation particularly.

No lasting tribute could therefore be more fitting than the Amy Johnson Memorial Scholarship for Women which it is now proposed to found. Amy Johnson's unique reputation will thus in death be the means of giving practical effect to her views, as she was perhaps never quite able to do in life.

~~~~~~~~~~~~~~~~~

AG Comments:

Again, another author and close friend made a marked contrast between the outer and inner persona of Amy. This anonymous writer puts her finger on the (two) secrets of Amy's global popularity: (1) she escaped the treadmill of ordinary life by her own effort; and (2) she was not "out for what she could get"...yes, "she did reap big rewards" and enjoyed them, but they were incidental.

Speculating, it is possible that Amy will have put much of her good fortune down to Lady Luck. It was also commented that "she did some silly things at times" – perhaps a little like the 'three-day millionaires' of Hessle Road when home between trips and spending their money like it was going out of fashion (Gill. 2010b).

I would add to the above list of "things which Amy did want were friends, flying and fun" – a major fourth factor: FREEDOM. That is, especially freedom from social expectations placed upon her by others and especially the suffocating restrictions imposed upon her for being 'female'. Amy was ahead of her time in rebelling against unladylike activities such as riding a bicycle or flying a plane. Her rebellious, questioning nature, I believe, was forged during her Hessle Road upbringing.

Even during her Eversleigh House Private School recollections, Amy stated "It was, perhaps, my tomboy spirit which, thank God, saved me from becoming a blue-stocking" (p135). This is why I am keen to stress and portray Amy as a "Hessle Road Tomboy". This 'badge' enabled her to revolt against being labelled as "ladylike" and fuelled her urge to gain freedom from sexist stereotypical labels.

Ending a book is not always easy; but I decided to take a leaf out of The Woman Engineer Journal. They ended their Memoriam by reprinting extracts from some of Amy's speeches when she was President of the Women's Engineering Society. Therefore, I am happy to let Amy have the final word in the next, final Appendix.

##################

(15J) Extracts from Speeches by Amy Johnson

Appendix 15J: JOHNSON, Amy. March 1941. Extracts. Speeches by Miss Johnson to Woman Engineer: the Organ of the Women's Engineering Society. March 1941. Journal. Amy Johnson – In Memoriam. 5. 6. 86.

"My flight was carried out for two reasons: because I wished to carve for myself a career in aviation, and because of my innate love of adventure." March, 1931.

"We women are just now on the threshold of another career which has so far been regarded as the strict province of man — that of aeronautical engineering...

"The only argument that men can bring forward against woman's intrusion is that of physical strength, but this seems to me very poor grounds for establishing and retaining a monopoly. After all physical strength is purely relative — there are some women stronger than some men. In engineering there are many jobs beyond a man's strength. What does he do? He fetches an instrument. What did I do when I found a job beyond my strength? At first I used to fetch a real man engineer, and if he couldn't do the job he'd fetch some tool that would. I soon learned that it saved time to fetch the tool right away.

"Women, I am sure, share with men the vital qualities needed in aeronautical engineering — patience, skill, delicate fingers, and a fertile mind. There is surely no reason whatever why we should not make good, whether it be in the design department, the workshops, or the repair shops. Anyhow, we're going to try." March, 1932.

"Progress in aviation, as in every sphere, is due to the people who believe nothing to be impossible. The course of ease is to say it cannot be done. The sceptics actually do much to further progress — they hold the pistol at the head of the dreamer and the optimist, challenging them to bring their dreams to reality. The answer of Progress to 'It can't be done' is 'Hold tight and watch'. Progress means always trying to go one better. To get more out of something, by trial and error, by crying for the moon, by hitching one's waggon to a star, by never saying die, so the world progresses. Individual steps may be too small to matter much, but the sum total, like the ant heap, the honeycomb, the skyscraper, or the 'Queen Mary' is an achievement worthwhile.

"In 'Modern Times', Charlie Chaplin's picture, there is portrayed a time when man succumbs to the machine. I should hate to imagine such a possibility, and frankly, I think it highly improbable. Behind the machine there is the mind, creator and controller of the machine. Human emotions will always rise superior to any degree of mechanisation, and we must retain the machine as a servant to do 'the chores' of life, leaving us, freedom for leisure, pleasure and High Thought." September, 1936.

"Why is it that there are not many more women employed in aviation? I believe that the fault lies as much with the women aspirants themselves as with the employer of labour so often captioned as hard-hearted, prejudiced and unjust.

"It is a significant fact that in every case of a woman achieving success, she is hailed by her firm as a real treasure and has become part and parcel of the organisation.

"As a general rule it may be admitted that technical efficiency is not the only qualification for a job. Book-learning and skill can be acquired by every eight out of ten merely by patient application to one's instructors. About 15% of real success is due to technical efficiency, and about 85% to skill in human engineering — to personality and the ability to deal with people.

"I would say that women have such a struggle and uphill fight that by the time they have acquired the technical skill equal to a man's, they have acquired something a great deal more valuable and of vast potential importance to their future employer — personality.

"To women who may sometimes feel they are not being given their dues I would like to say this: We should try not to start off in a spirit of resentfulness and aggression. Sometimes we are our own worst enemies. We argue and try to convince that we are just as good as any man and we are amazed when our belligerent tactics go unrewarded, when we fail to get the job and complain bitterly of inequality and injustice.

"Instead we should be first of all sure of ourselves on the technical side of the job and spend the rest of our energies putting ourselves over. Women are noted for talking. Well, remember that it is said that leadership gravitates to the man who can talk. Lowell Thomas once said in a speech — and how truly — that the man who can speak acceptably is usually given credit for an ability out of all proportion to what he possesses.

"You need not wait for an after dinner speech to try this out — try it at interviews." September, 1937.

"Believe nothing to be impossible"

Amy Johnson (1936).

(16) ACKNOWLEDGEMENTS: With Thanks

Michéle Beadle (manuscript reader); Carnegie Heritage Centre staff; Judy Chilvers; Steve Clarke; Michelle Coldham; Albert Kenneth Davis; Peter Devitt; Gerry Duffy; John & Kay Dunn; Audrey Dunne (manuscript reader); Peter Elliott; Ian Gill; Tom Goulder; Dianne Hargreaves; Brian Hodgins; Hull History Centre staff; Ernie Hunter; John & Fiona Kelly; Kevin Marshall; Jim Mitchell; Richard Nielsen (manuscript reader); Keith Parker; Kath Peterson; Christine Pinder; Andrae Sutherland (manuscript reader); Rick Welton; Rob Wilson; and Angus Young.

(17) BIBLIOGRAPHY:
My Own C-T-S Referencing Style
See full details at http://academicreflexions.blogspot.co.uk/

ANDREW JOHNSON KNUDTZON LTD. 2015. Website. Welcome to AJK. http://www.ajkltd.co.uk/index.asp. Accessed 3 December 2014.

ANONYMOUS. March 1941. Obituary. Amy by a fellow pilot of the ATA. Woman Engineer: the Organ of the Women's Engineering Society. March 1941. Journal. Amy Johnson – In Memoriam. 5. 6. 85. See Appendix 15I.

BAILEY, Eva. 1987. Book. Amy Johnson. London. Hamish Hamilton Profile.

BANNER, Herbert S. 1933. Book. Amy Johnson. London. Rich & Cowan. Popular Lives Series.

BASSET HOUND TRIO. 11 December 2013. YouTube. Amy, Wonderful Amy – song and lyrics. Jack Hylton and His Orchestra. Victrola Credenza. https://www.youtube.com/watch?v=OGQJx39Ygh4. Accessed 3 March 2015.

BBC TV. 21 October 2002. Documentary. Amy Johnson. Inside Out – Yorkshire & Lincolnshire. http://www.bbc.co.uk/insideout/yorkslincs/series1/amy-johnson.shtml. Accessed 4 December 2014.

BBC TV. 8 September 2010. Documentary. Spitfire Women: ATA Female Pilots. iPlayer. Accessed 12 July 2015.

CHESTER, Donald. 1980. Book. Silvered Wings: A Commemorative Brochure to Mark the 50th Anniversary of Amy Johnson's Solo Flight to Australia. Hull. City Council.

CLYDE BUILT SHIPS DATABASE. 2015. Website. Steamship Naldera. http://www.clydesite.co.uk/clydebuilt/viewship.asp?id=15358. Accessed 4 December 2014.

COLTMAN STREET VILLAGE PROJECT. 2015. Website. House-by-House Directory of Coltman Street. http://www.coltmanstreet.co.uk/directory.shtml. Initiated by Angie Storr. 26. Accessed 19 March 2003.

COOLING, Rupert. 5 January 1991. Article. Amy: Mystery Still Unsolved. Hull Daily Mail. Feature.

CRAWLEY, Paul. 1989. Book. The Ninth Parachute: Amy Johnson and the Mystery of Mr X. Hull. Malet Lambert Series.

EDWARDS & LOCKETT'S POTTERY. 2012. Website. Edwardian Bespoke Decorated Fine Bone China & Earthenware. http://www.edwardianchina.co.uk/index.html. Accessed 9 February 2015.

ELSOM, Ken. 1990. Book. Hull Personalities: Pearson Park and Avenues Area. Hull. The Avenues Press.

FINCH, Bob. 1989. Book. Amy Johnson: Global Adventurer. Hull. College of Further Education. Local History Archive Unit Publication.

GILL, Alec & SARGEANT, Gary. 1985. Book. Village Within a City: The Hessle Road Fishing Community of Hull. Hull. University Press.

GILL, Alec. 1989. Book. Lost Trawlers of Hull: Nine Hundred Losses between 1835 and 1987. Beverley. Hutton Press.

GILL, Alec. 2003. Book. Hull's Fishing Heritage: Aspects of Life in the Hessle Road Fishing Community. Barnsley. Pen and Sword Books.

GILL, Alec. 2010a. DvD. Hull Fish Dock: St. Andrew's - One Big Family. Hull. Wordspin Productions. Amazon. http://www.amazon.co.uk/dp/B00BCYNPIM. Accessed 28 May 2015.

GILL, Alec. 2010b. DVD. Three-day Millionaires: Hull Trawlermen Home Between Trips. Hull. Wordspin Productions. Amazon. http://www.amazon.co.uk/dp/B00BCXU4QE. Accessed 24 March 2015.

GILL, Alec. 2010c. DVD. Arctic Trawlermen: Last of the Hunters. Hull. Wordspin Productions. Amazon. http://www.amazon.co.uk/dp/B00BCMG27A. Accessed 29 June 2015.

GILL, Alec. 2013. eBook. Superstitions: Folk Magic in Hull's Fishing Community. Amazon. Kindle Direct Publishing. https://www.amazon.co.uk/dp/B00EWOBIJM. Accessed 24 March 2015.

GILLIES, Midge. 2003. Book. Amy Johnson: Queen of the Air. London. Weidenfeld & Nicolson.

GOWER, Pauline. 8 January 1941. Letter. The Head of the Women's Section ATA. The (London) Times. Reproduced in The Woman Engineer: the Organ of the Women's Engineering Society. March 1941. Journal. Amy Johnson – In Memoriam. 5. 6. 84. See Appendix 15H.

GREY, Elizabeth. 1966. Book. Winged Victory: The Story of Amy Johnson. London. Constable Young Books Ltd.

HARADA, Yoshifumi & HUNTER, Murray. 28 March 2013. Website. Walking under a ladder: Superstition and ritual as a cognitive bias in management decision making. The Nordic Page. TNP Global Outlook. http://www.tnp.no/norway/global/3635-walking-under-a-ladder-superstition. Accessed 5 August 2015.

HARRISON, Amanda J. 2015. Website. Female Commercial Tiger Moth Pilot. http://amandajharrison.com/. Accessed 25 February 2015.

HASLETT, Caroline. 14 January 1941. Obituary. Letter to the Times. Reproduced in Woman Engineer: the Organ of the Women's Engineering Society. March 1941. Journal. Amy Johnson – In Memoriam. 5. 6. 84. See Appendix 15G.

HELDERLINE. 2015. Website. Oil Tanker Phorus. http://www.helderline.nl/tanker/1410/phorus/. Accessed 21 June 2015.

HULL DAILY MAIL. 9 October 1973. Newspaper. At Amy's First School. Women. Miss Humber. 6. See Appendix 15B.

HULL HISTORY CENTRE. 15 December 1922. Website. Amy Johnson's Letters. Reference L.DIAJ/7. 4 of 40. http://catalogue.hullhistorycentre.org.uk/files/l-diaj.pdf. Accessed 15 June 2015.

JOHNSON, Amy. April 1921. Journal. New Year's Day in China. The Boulevardian. 21. 2. 27-28. Available via The Old Kingstonians' Association at the Carnegie Heritage Centre. See Appendix 15C.

JOHNSON, Amy. 1938. Chapter. Amy Johnson. Myself When Young by Famous Women of Today. Edited. The Countess of Oxford and Asquith. London. Frederick Muller Ltd. 131-156. See Appendix 15A.

JOHNSON, Amy. 1939. Book. Sky Roads of the World. London. W & R Chambers. See Appendix 15D.

JOHNSON, Amy. 1 January 1941. Chapter + Letter. A Day's Work in the ATA. Woman Engineer: the Organ of the Women's Engineering Society. March 1941. Journal. Amy Johnson – In Memoriam. 5. 6. 88-90. See also Appendices 15E, 15F, 15G, 15H, 15I, and 15J Extracts of Speeches).

LENTON, H.T. & COLLEDGE, J.J. 1973. Book. Warships of World War II. 2nd Edition. London. Ian Allan.

LONDON GAZETTE. 16 May 1941. 2114 Honours and Awards Supplement. 2.

LUFF, David. 2002. Book. Amy Johnson: Enigma in the Sky – An Official Biography. Shrewsbury. Airlife Publishing.

METHODIST CONFERENCE. 12-23 July 1938. Report Handbook. Methodism in Hull. Held at Thornton Hall, Great Thornton Street, Hull. A. Brown & Sons.

PETRE, Jonathan. 4 July 2015. Web Article. Amy flies again…repeated 85 years on by bold aviatrix. The Mail on Sunday. 1-8. http://www.dailymail.co.uk/news/article-3149712/Amy-flies-solo-global-flight-tragic-death-inspired-world-Amy-Johnson-s-amazing-journey-repeated-85-years-bold-aviatrix-MoS-backing-just-that.html. Accessed 14 July 2015.

PRE-WAR CAR. 2011. Website. The Fruits of Flying. http://www.prewarcar.com/magazine/previous-features/the-fruits-of-flying-015752.html. Accessed 2 March 2015.

PRISTON, Jane. 2015. Website. Amy Johnson & Herne Bay Project. http://www.amyjohnsonhernebay.com/. Accessed 27 January 2015.

SHAKESPEARE, William. 1601. Play. Hamlet. Act 1. Scene 3. 78-82. http://www.enotes.com/shakespeare-quotes/thine-own-self-true. Accessed 9 May 2015.

SMITH, Constance Babington. 1967. Book. Amy Johnson. London. Collins.

SNELL, Gordon. 1980. Book. Amy Johnson: Queen of the Air. London. Hodder & Stoughton.

THOMPSON, Michael. 1989. Book. Fish Dock: The Story of St.Andrew's Dock, Hull. Beverley. Hutton Press.

VINCENT, John. 16 November 1996. Article. Why Amy Johnson took flight to escape fame. The Times. 13.

WELTON, Rick. 2015. FaceBook Group. Amy Johnson Festival 2016. https://www.facebook.com/amyjohnson2016?pnref=story. Accessed 13 May 2015.

WOMAN ENGINEER: the Organ of the Women's Engineering Society. March 1941. Journal. Amy Johnson – In Memoriam. 5. 6. 81-90.

WOOD, Alexandra. 24 May 2005. Article. Pioneer 'foolhardy' for historic air-trip. Yorkshire Post. 3.

YORKSHIRE AIR MUSEUM. 20 March 2013. Website. Recalling How Halifax Friday the 13th Got its Name. Ian Richardson. http://www.yorkshireairmuseum.org/air-museum-news/recalling-how-halifax-friday-the-13th-got-its-name. Accessed 14 July 2015.

===============================

(18) DEDICATION: Appreciation

All the former Cod Farm and Fish Dock workers, especially my good friend and Hessle Roader Ernie Hunter (1937-2015)
– plus Jack Anderson and Albert Davis.

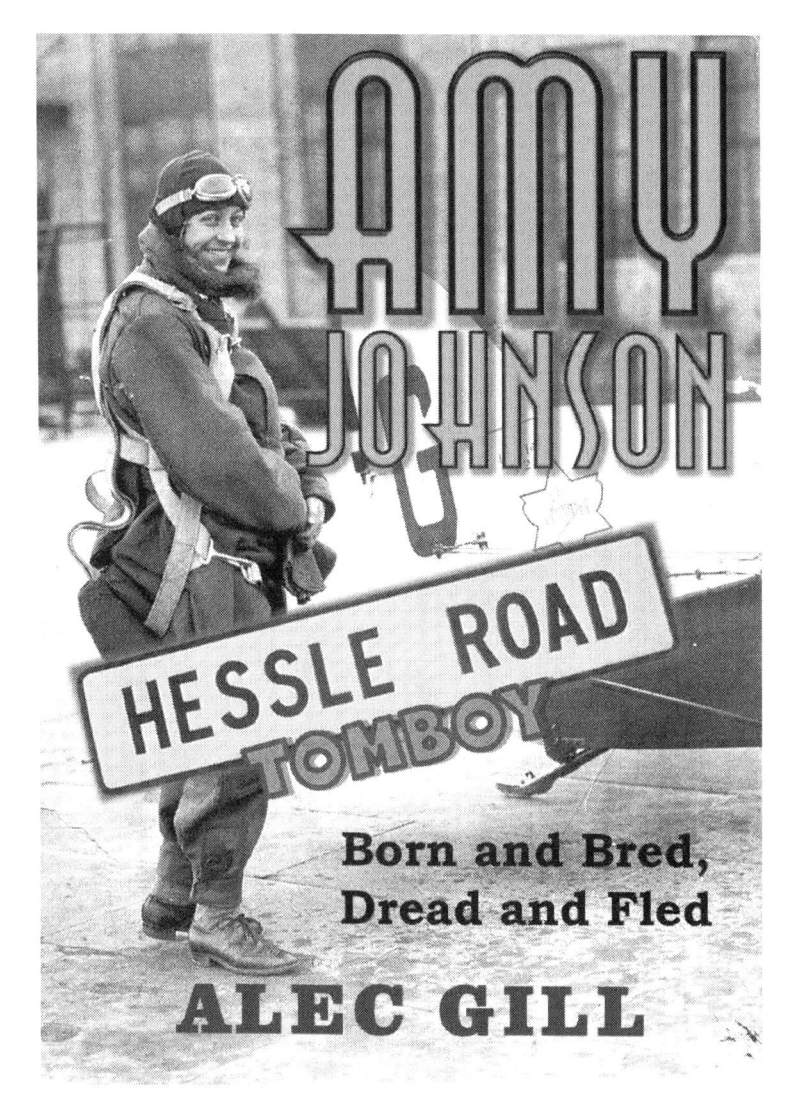

AMY JOHNSON
HESSLE ROAD TOMBOY
Born and Bred, Dread and Fled
ALEC GILL

Image 60: TOMBOY SMILE: This is a happy photograph of Amy Johnson about to set off on her solo flight to Australia. She has the loveliest of smiles - plus a twinkle in her eye. In many pictures, she looks so serious and, some with Jim Mollison, are very unhappy. So this is a refreshing image. Courtesy Hull Daily Mail – with much appreciation.

PLEASE NOTE: There are also two other versions of this same book by me:
(1) a printed Colour version – much more expensive – copy; and
(2) an eBook Amazon Kindle version for eReaders, iPads, etc:
https://www.amazon.co.uk/AMY-JOHNSON-Community-Trawling-Heritage-ebook/dp/B015ST0GEM.

(19) ABOUT: The Author and Book

I love Hessle Road, its culture, heritage and, especially, its people – the fishing families of Hull (Yorkshire, England). My research into the community began back in 1974 as a social documentary photographer. This work resulted in twenty solo exhibitions in both the UK and the USA. By the mid-1980s, I moved on to become an author/researcher of the fishing community. This resulted in six printed books, one of which sold over 10,000 copies – not bad for a local publication. Come the mid-1990s, I was ready for another change of direction – but not topic. I moved into film-making with the production of six VHS commercial videos with Dovedale Studios in Hull. They too proved very popular.

Alongside this work, I presented countless talks and published scores of articles. I enjoy public speaking and audience participation – their views and information are priceless. In addition, there have been numerous and regular radio and TV appearances on both sides of the Atlantic.

During 2002-2012, I took up an academic post at the University of Hull. During this period, I was honoured with an MBE at Buckingham Palace (2009) and awarded an Honorary Doctorate by the University of Hull (2010) for all my Hessle Road Fishing Community research and campaign work (especially for jointly initiating the annual and on-going Lost Trawlermen's Day open-air memorial service). Although I did not have to retire from the world of academia, I was eager to return full-time to all my Hessle Road writing once more – and did so in October 2012.

Since then, I have converted all six VHS video tapes into DvDs (now available via Amazon). In addition, I began to write eBooks by using the Amazon/Kindle global platforms – AMY is my fourth such publication – and many more are planned. I also intend to return to my 1974 B&W photo roots. Thanks to top photographer and BAFTA award-winning film-maker Paul Berriff OBE, there are now moves to get many of my 6,636 images converted into digital formats (from the old-fashioned Rolleicord 120 rollfilm 6x6cm negatives). Given this cyber basis, there are plans for a coffee-table book of my best 500 Hessle Road images. This publication will coincide with exhibitions and talks around the world to global audiences – and perhaps dovetailing with Hull's City of Culture during 2017.

I have wanted to produce a detailed eBook for some time. Fortunately, its release (October 2015) coincided with the upcoming celebration of the Amy Johnson Festival 2016 – a year-long event to mark 75 years since her death in the Second World War. Events are not only in Hull, but also around the globe. Since releasing the eBook, numerous people declared their strong preference for a printed version of Amy – so here it is in Summer 2016. Enjoy – I hope.

I hope that my local history research into Amy's childhood background and formative years adds to a deeper understanding and appreciation of this incredible woman. I was keen to highlight her ancestry from a Danish immigrant; how her grandfather Andrew Johnson did so much to build up Hull's Victorian trawling industry; and the significance of the Cod Farm and its workers upon the Johnson Family's prosperity. In addition, how Amy herself grew up in the streets of the Hessle Road Fishing Community, her education at the Boulevard School and how the local folklore beliefs shaped her adult personality.

As a psychologist, I was delighted to explore Amy's dread-and-fled and hermit traits. I believe that her early experience accounts for why she became rootless and restless until she found her tomboy niche as a pilot in the air and an engineer in the maintenance workshops. Aged 26-years old, she finally found her forte, acquired her open cockpit plane (thanks primarily to her father's sponsorship) and flew to Australia. Her high level of self-determination also played a significant part in this triumph.

I hope this book firmly links Amy with her true roots within Hull's Hessle Road Fishing Community. For whatever reason(s), it seems to me that Amy was keen to conceal her place of birth. She was extremely successful as "a most fertile liar" and, as a result, distorted history. I hope that my work enriches our perspective of Hull's hero, Amy Johnson.

Image 61: Author Dr. Alec Gill MBE on Hull Pier by the Humber Estuary - with The Deep in the background of this night-time shot (Feb 2009). Amy Johnson is my fourth eBook in a long-term series (and now this printed version – thanks to Amazon's CreateSpace). I plan to produce a variety of eBooks drawing upon my Hessle Road research and writing which began in 1974. Copyright Audrey Dunne.

(20) ENDNOTES: Sources and Snippets

[1] BASSET HOUND TRIO. 11 December 2013. YouTube. Amy, Wonderful Amy – song and lyrics. Jack Hylton and His Orchestra. Victrola Credenza. https://www.youtube.com/watch?v=OGQJx39Ygh4. Accessed 3 March 2015.

[2] JOHNSON, Amy. 1938. Chapter. Amy Johnson. Myself When Young by Famous Women of Today. Edited. The Countess of Oxford and Asquith. London. Frederick Muller Ltd. 139. See Appendix 15A.

[3] EDWARDS & LOCKETT'S POTTERY. 2012. Website. Edwardian Bespoke Decorated Fine Bone China & Earthenware. http://www.edwardianchina.co.uk/index.html. Accessed 9 February 2015. They are a friendly, family-run firm. I visited many times, re-stocked lots of Hessle Road plates, and always received a warm welcome.

[4] LUFF, David. 2002. Book. Amy Johnson: Enigma in the Sky – An Official Biography. Shrewsbury. Airlife Publishing. 36.

[5] LUFF. 2002. 110. Luff, however, did not cite his newspaper source, but it was probably the London Evening News who reported that Amy came from a "wealthy Midlands family". It is well known that reporters 'get things wrong' and his article did so in a number of respects; but I just wonder if Amy deliberately left her place of birth as vague as possible to throw him (and readers) off the scent of her father's Hull-based business.

[6] ANDREW JOHNSON KNUDTZON (AJK) became a subsidiary of the Andrew Marr International (AMI) group around 1953. Marr are another family business with solid fishing roots in the ports of Hull and Fleetwood. If the Johnsons had had just one son, the company may well have continued as an independent family concern. The present-day AJK firm have a website that gives details of their business activities: http://www.ajkltd.co.uk/index.asp. Accessed 11 December 2014.

[7] Incidentally, it was on a 5th January (67-years later) in 1941 when their granddaughter Amy Johnson was killed in the Thames Estuary.

[8] HULL DAILY MAIL. 8 January 1924. 7.

[9] When Andrew Anglicised his name (1878), he and Mary lived at No.67 William Street in the fishing community. I am very grateful to Andrae Sutherland who kindly offered to research deeper into the Danish side of the Johnson Family. She did a marvellous job, especially when researching Census data. Andrae found that the Johnson Family lived at various Hessle Road addresses over the decades:
1881 = No.36 Great Thornton Street;
1891 = No.326 St.George's Road;
1901 = No.292 South Boulevard;
1911 = No.111 Coltman Street; and by
1924 (their Golden Wedding), they resided at No.509 Anlaby Road (near Humber St.Andrew's Club) – before they moved to Bridlington.

[10] COLTMAN STREET VILLAGE PROJECT website, by Angie Storr, clearly shows that Andrew Johnson lived at No.111 Coltman Street in 1910 - next door to another trawler owner Thomas Boyd at No.110 (not at the same time) – but I have yet to establish how long the large Johnson family were resident in this posh street. The website is: http://www.coltmanstreet.co.uk/directory.shtml. 26 of 49 (Accessed March 2003) – well worth a visit and a model of what any residents could do for their district. There was also a Holmes Family (perhaps Amy's Grandma Johnson's relatives) at No.179 Coltman Street – now a Grade II Listed Building.

[11] GILL, Alec. 1989. Book. Lost Trawlers of Hull: Nine Hundred Losses between 1835 and 1987. Beverley. Hutton Press. 59. In addition to Flower of the Forest, I have also traced details of Flower of the Valley (Port No.13/1885) that was sold to Denmark in May 1896. Other authors have mentioned a Flower of the Field and Flower of the Harvest, but I have so far been unable to trace these vessels in my handwritten records copied from H.M.Customs & Excise Shipping Registers from the 1830s onwards.

[12] Janet Scott attended one of my classes about Hull's Fishing Heritage at The University of Hull (January 2003). I am very grateful to her for the extra snippets of information, reading my 2003 chapter on Amy to double check the details, and lending me some photographs.

[13] Again, I am indebted to Andrae Sutherland for tracking down information about Andrew Johnson as owner and manager of the Anglo Norwegian

Fishing Company Limited. I had no previous idea that AJK and Anglo Norwegian trawlers were linked.

[14] GILLIES, Midge. 2003. Book. Amy Johnson: Queen of the Air. London. Weidenfeld & Nicolson. 9.

[15] GILLIES. 2003. 11.

[16] JOHNSON. 1938. 131.

[17] JOHNSON. 1938. 133-34.

[18] JOHNSON. 1938. 138.

[19] SMITH, Constance Babington. 1967. Book. Amy Johnson. London. Collins. 26.

[20] JOHNSON. 1938. 141.

[21] SMITH. 1967. 113.

[22] SMITH. 1967. 25; LUFF. 2002. 33; and GILLIES. 2003. 14.

[23] JOHNSON. 1938. 135.

[24] JOHNSON. 1938. 137.

[25] JOHNSON. 1938. 142.

[26] HULL DAILY MAIL. Thursday 11 September 1902. 5.

[27] METHODIST CONFERENCE. 12-23 July 1938. Report Handbook. Methodism in Hull. Held at Thornton Hall, Great Thornton Street, Hull. A. Brown & Sons. 17.

[28] JOHNSON. 1938. 132-33.

[29] GILLIES. 2003. 12-14.

[30] ST.GEORGE'S WESLEYAN METHODIST MINUTE BOOK (1910-1951) was rescued from a rubbish skip by Hull historian Chris Ketchell (October 1984). I was surprised to realise that the photocopies I made from his document were over thirty years old (January 2015). That, to me, is the appeal of doing long-term, dedicated research into one specific community. Eventually, some of the scattered pieces of information begin to slot together in a harmonious way. Sadly, Chris has since died and I have no idea if the Minute Book is still available anywhere.

[31] LUFF. 2002. 31.

[32] JOHNSON, Amy. April 1921. Journal. New Year's Day in China. The Boulevardian. 21. 2. 27-28. Available via the Old Kingstonians' Association at the Carnegie Heritage Centre. See Appendix 15C.

[33] LUFF. 2002. 82.

[34] GILLIES. 2003. 327.

[35] LUFF. 2002. 37.

[36] LUFF. 2002. 28.

[37] HULL DAILY MAIL. 9 October 1973. Newspaper. At Amy's First School. Women. Miss Humber. 6. See Appendix 15B.
[38] JOHNSON. 1939. 135.
[39] JOHNSON. 1938. 137.
[40] JOHNSON. 1938. 137.
[41] JOHNSON. 1938. 139.
[42] GREY, Elizabeth. 1966. Book. Winged Victory: The Story of Amy Johnson. London. Constable Young Books Ltd. 3.
[43] JOHNSON. 1938. 137.
[44] JOHNSON. 1938. 138.
[45] JOHNSON. 1938. 138.
[46] JOHNSON. 1938. 140-41.
[47] SMITH. 1967. 24.
[48] GILLIES. 2003. 18.
[49] FINCH, Bob. 1989. Book. Amy Johnson: Global Adventurer. Hull. College of Further Education. Local History Archive Unit Publication. 53 and 59.
[50] GREY. 1966. 42.
[51] JOHNSON. 1938. 140.
[52] JOHNSON. 1938. 139.
[53] THE AVENUES RESIDENTS' ASSOCIATION has, quite rightly, made much of the fact that Amy Johnson once lived in their neighbourhood. A book was published about the famous people who stayed in this leafy middle-class part of the city. ELSOM, Ken. 1990. Book. Hull Personalities: Pearson Park and Avenues Area. Hull. The Avenues Press. 44. But, in reality, Amy herself only resided in Park Avenue for a mere four years of her life (1918-22) prior to leaving for Sheffield University. After graduation, she only worked in Hull for a matter of months – and even then moved into a flat on Anlaby Road. Mother Ciss refused to let newly-graduated Amy use the family home when it was empty for three months during the summer of 1925. The Amy-Avenues link, therefore, was not profound – and had very little impact in shaping her personality. Amy's parents were still living in Park Avenue when she made her solo flight to Australia (1930). Thus, the association between Amy and The Avenues was forged in the mind of the Hull public and the world at large. It was falsely assumed that she must have grown up in that posh area too – and nowhere near Hull's fishing community.
[54] SMITH. 1967. 27.

[55] SMITH. 1967. 28.

[56] SMITH. 1967. 273. In 1931, Amy's own parents also retired to live in Bridlington – granddad Andrew Johnson had retired there much earlier. Amy, subsequently, performed the opening ceremony of Sewerby Hall (north of Bridlington) on 1 June 1936. In 1959, father Will Johnson presented Amy's trophies and artefacts to the Borough of Bridlington. Sewerby Hall now has over 130 items in the Amy Johnson Collection. The venue is well worth a visit. Check their website http://www.sewerbyhall.co.uk/hall/exhibitions/.
I donated a copy of my Women of Hessle Road plate to them – I wonder if it was ever displayed.

[57] THOMPSON, Michael. 1989. Book. Fish Dock: The Story of St.Andrew's Dock, Hull. Beverley. Hutton Press. 18.

[58] GILL, Alec. 2010a. DvD. Hull Fish Dock: St.Andrew's - One Big Family. Hull. Wordspin Productions. Amazon. http://www.amazon.co.uk/dp/B00BCYNPIM. Ch.12. 44:22. Accessed 28 May 2015.

[59] Albert Kenneth Davis and I became acquainted through various Hull fishing FaceBook groups during 2014/15. He lives away from Hull and we have never met – but that is the wonder of our worldwide web these days. He provided some priceless material about how Cod Farm worked – the good and bad side of the business.

[60] Hull Fish Meal & Oil Company Limited (HFM) was established in 1891 by a consortium of trawler owners and fish merchants (no doubt including AJK). The giant HFM factory was located on the south side of St.Andrew's Fish Dock – not too far from Cod Farm. In the 1930s, the plant had six processing units that converted slimy, smelly offal / waste and rotten fish into sterilised, dry animal feedstuff. Annually, this amounted to around 35,000 tons. Their successful Provimi 66 supplemented the diet of poultry and cattle. Other by-products, from fish oils, were used in the production of leather, steel, soap, margarine, and glue products. The HFM factory worked 24/7 all year round.

[61] NHS Choices. 12 December 2014. Website. Symptoms of Leptospirosis. http://www.nhs.uk/Conditions/Leptospirosis/Pages/Symptoms.aspx. Accessed 7 September 2015.

[62] GILL. 2010a. Ch.12. 44:20.

[63] SHAKESPEARE, William. 1601. Play. Hamlet. Act 1. Scene 3. 78-82. http://www.enotes.com/shakespeare-quotes/thine-own-self-true. Accessed 9 May 2015. This quote is from Polonius giving worldly advice to his departing son Laertes – not that Polonius was a paragon of virtue or practiced what he preached.

[64] JOHNSON. 1938. 143.
[65] JOHNSON. 1938. 142.
[66] FINCH. 1989. 87. I suspect a family member found this teaching job for Amy; but I have no proof.
[67] GILLIES. 2003. 24.
[68] JOHNSON. 1938. 144-45.
[69] LUFF. 2002. 54-55.
[70] JOHNSON. 1938. 147.
[71] JOHNSON. 1938. 147.
[72] GILLIES. 2003. 34-36.
[73] GILLIES. 2003. 33-39. Midge Gillies entitled her Chapter Three, Sibling Rivalry, and compared Amy with Irene in various ways.
[74] GILLIES. 2003. 24.
[75] HULL HISTORY CENTRE. 15 December 1922. Website. Amy Johnson's Letters. Reference L.DIAJ/7. 4 of 40. http://catalogue.hullhistorycentre.org.uk/files/l-diaj.pdf. Accessed 15 June 2015.
[76] GILLIES. 2003. 28.
[77] GILLIES. 2003. 35.
[78] GILLIES. 2003. 28.
[79] GILLIES. 2003. 49.
[80] JOHNSON. 1938. 137, 141, 150, 151, 153 (twice).
[81] JOHNSON. 1938. 150-51.
[82] JOHNSON. 1938. 152-53.
[83] GILLIES. 2003. 61-62.
[84] JOHNSON. 1938. 154.
[85] LUFF. 2002. 134-47. His Chapter 9 is entitled Call Me Johnnie.
[86] JOHNSON. 1938. 134.
[87] GILLIES. 2003. 91.
[88] BAILEY, Eva. 1987. Book. Amy Johnson. London. Hamish Hamilton Profile. 20.
[89] FINCH. 1989. 25.
[90] LUFF. 2002. 105.
[91] BAILEY. 1987. 25.
[92] GILLIES. 2003. 109.

[93] GILL, Alec. 2013. eBook. Superstitions: Folk Magic in Hull's Fishing Community. Amazon. Kindle Direct Publishing. https://www.amazon.co.uk/dp/B00EWOBIJM. Accessed 24 March 2015. Chapter 23. Forbidden Colour.

[94] SMITH. 1967. 182.

[95] GILLIES. 2003. 98.

[96] YORKSHIRE AIR MUSEUM. 20 March 2013. Website. Recalling How Halifax Friday the 13th Got its Name. Ian Richardson. http://www.yorkshireairmuseum.org/air-museum-news/recalling-how-halifax-friday-the-13th-got-its-name. Accessed 14 July 2015. The Museum have reconstructed a full-scale bomber named "Friday the 13th" in honour of Halifax, LV907, which completed 128 operations with 158 Squadron - it is representative of all Halifax bombers.

[97] Olive Westcott was at my talk (Our Grannies Superstitions) to the Ideal Standard Retired Group in October 1991 and she gave me this positive version of the negative Whistling Woman superstition against women.

[98] JOHNSON. 1938. 142.

[99] BANNER, Herbert S. 1933. Book. Amy Johnson. London. Rich & Cowan. Popular Lives Series. 29.

[100] SMITH, 1967. 181-2. The nearest I have come to finding Jason's Kippers was in an old advertisement (undated and unsourced). At each corner of the advert are four oblong shapes. In each is printed the following four AJK brand names: Trawl JASON Fish; Norway JASON Herrings; Large JASON Prawns; and Prime JASON Kippers. The quality of the photocopy is so poor that it was not worth trying to reproduce it – but still searching for a better copy and/or an actual label.

[101] SMITH. 1967. 285-86.

[102] HARADA, Yoshifumi & HUNTER, Murray. 28 March 2013. Website. Walking under a ladder: Superstition and ritual as a cognitive bias in management decision making. The Nordic Page. TNP Global Outlook. http://www.tnp.no/norway/global/3635-walking-under-a-ladder-superstition. Accessed 5 August 2015.

[103] BAILEY. 1987. 26.

[104] SMITH. 1967. 191.

[105] FINCH. 1989. 61.

[106] SMITH. 1967. 191.

[107] CHESTER, Donald. 1980. Book. Silvered Wings: A Commemorative Brochure to Mark the 50th Anniversary of Amy Johnson's Solo Flight to Australia. Hull. City Council. 9.

[108] GILLIES. 2003. 123.

[109] WOOD, Alexandra. 24 May 2005. Article. Pioneer 'foolhardy' for historic air-trip. Yorkshire Post. 3.

[110] SNELL. 1980. 79.

[111] BAILEY. 1987. 43.

[112] SNELL. 1980. 82.

[113] CLYDE BUILT SHIPS DATABASE. 2015. Website. Steamship Naldera. http://www.clydesite.co.uk/clydebuilt/viewship.asp?id=15358. Accessed 4 December 2014.

[114] GILLIES. 2003. 178.

[115] Jack Anderson was a student in my I Remember When: Reminiscence Writing Class at The University of Hull Centre for Lifelong Learning. He was a natural storyteller and often had a twinkle in his eye with some of his tales. I taught these classes for several years until I resigned in 2002 to work for the Study Advice Service at the University - on a more regular basis than as a part-time lecturer. These became ten very happy years working with Katy Barnett who ran the team of advisors (2002-2012). I then retired from academia to return to my writing career.

[116] PreWarCar. 2011. Website. The Fruits of Flying. http://www.prewarcar.com/magazine/previous-features/the-fruits-of-flying-015752.html. Accessed 2 March 2015.

[117] LUFF. 2002. 188-9. Hull Times. 30 January 1932. Photograph. At Amy's Old School. 11. The image shows the plaque and five people. Amongst the dignitaries is Amy's mother Ciss – the happiest looking person at the event.

[118] LUFF. 2002. 295.

[119] FINCH. 1989. 62.

[120] GILL, Alec & SARGEANT, Gary. 1985. Book. Village Within a City: The Hessle Road Fishing Community of Hull. Hull. University Press.

[121] SNELL. 1980. 100.

[122] GILLIES. 2003. 214.

[123] VINCENT, John. 16 November 1996. Article. Why Amy Johnson took flight to escape fame. The Times. 13.

[124] SMITH. 1967. 348.

[125] GILLIES. 2003. 327.

[126] GILLIES. 2003. 293-94.

[127] GILL, Alec. 2010b. DVD. Three-day Millionaires: Hull Trawlermen Home Between Trips. Hull. Wordspin Productions. Amazon. http://www.amazon.co.uk/dp/B00BCXU4QE. Accessed 24 March 2015.

[128] JOHNSON, Amy. 1939. Book. Sky Roads of the World. London. W & R Chambers. See Appendix 15D.

[129] GILLIES. 2003. 252.

[130] GILL, Alec. 2010c. DVD. Arctic Trawlermen: Last of the Hunters. Hull. Wordspin Productions. Amazon. http://www.amazon.co.uk/dp/B00BCMG27A. Accessed 29 June 2015.

[131] GILL. 1989. 93-99.

[132] Amy and marriage were incompatible. Her relationship with Jim Mollison was on the slide from early days and did not even reach the 'seven-year itch'. She started divorce proceedings in March 1937 and, around February 1938, it was granted. She reverted to her maiden name as soon as possible and insisted everyone call her Miss Amy Johnson again. I did not give too much attention to Jim Mollison in my story. He rarely came to Hull and helped Amy distance herself further from her Hull roots.

[133] JOHNSON. 1939.184-85.

[134] SMITH. 1967. 30. Happy Endings. 347-64. Her chapter gives a good overall account of Amy's final flight and plunge into the Thames.

[135] Women were not, during the Second World War, allowed to fly in combat – but did a tremendous job behind the scenes. Amy was never a politically active feminist, but spoke out against sexism when it affected her directly – especially with regard to inequality in pay between men and women in the forces. Amy was paid £6 per week.

[136] CRAWLEY, Paul. 1989. Book. The Ninth Parachute: Amy Johnson and the Mystery of Mr X. Hull. Malet Lambert Series. 6.

[137] WOMAN ENGINEER: the Organ of the Women's Engineering Society. March 1941. Journal. Amy Johnson – In Memoriam. 5. 6. 81-90. See Appendices E to J.

[138] WOMAN ENGINEER. 1941. 82.

[139] Some documents state Amy joined the ATA on 20 May 1940 and another gave October of that year (according to The Woman Engineer Journal p.83). If the latter is true, then she was only in the ATA for a few months. Due to male prejudice within the RAF hierarchy, only eight women were allowed to recruit during the Spring of 1940 – and Amy was not one of those. After Dunkirk and the Battle of Britain, enlistment of women into the ATA increased.

[140] BBC TV. 8 September 2010. Documentary. Spitfire Women: ATA Female Pilots. iPlayer. Accessed 12 July 2015.

[141] GILLIES. 2003. 328.
[142] CRAWLEY. 1989. 13.
[143] SMITH. 1967. 347.
[144] COOLING, Rupert. 5 January 1991. Article. Amy: Mystery Still Unsolved. Hull Daily Mail. Feature. 14.
[145] GILL. 2013. 24.
[146] LUFF. 2002. 343.
[147] SNELL. 1980. 115-16. GILLIES. 2003. 342.
[148] JOHNSON, Amy. 1 January 1941. Chapter. A Day's Work in the ATA. Woman Engineer: the Organ of the Women's Engineering Society. March 1941. Journal. Amy Johnson – In Memoriam. 5. 6. 89. See Appendices 15E and 15F.
[149] GOWER, Pauline. 8 January 1941. Letter. The Head of the Women's Section ATA. The (London) Times. Reproduced in The Woman Engineer: the Organ of the Women's Engineering Society. March 1941. Journal. Amy Johnson – In Memoriam. 5. 6. 84. See Appendix 15H.
[150] CRAWLEY. 1989. 14.
[151] BBC TV. 2010.
[152] I suspect that there was a serious mix up in the public mind with Amy's first lover Hans Arregger and his Germanic sounding name.
[153] LUFF. 2002. 339-41.
[154] GILLIES. 2003. 335.
[155] LENTON, H.T. & COLLEDGE, J.J. 1973. Book. Warships of World War II. 2nd edition. London. Ian Allan. 310.
[156] CRAWLEY. 1989. 26.
[157] LUFF. 2002. 328.
[158] Harry Gould Junior approached me after I had presented a talk at The Kingston General Hospital Stroke Group (1991) to tell me about his father and the death of Amy Johnson in the Thames Estuary. I made notes and visited him later to borrow photographs of his dad.
[159] SMITH. 1967. 360.
[160] BBC TV. 21 October 2002. Documentary. Amy Johnson. Inside Out – Yorkshire & Lincolnshire. http://www.bbc.co.uk/insideout/yorkslincs/series1/amy-johnson.shtml. Accessed 4 December 2014.
[161] GILLIES. 2003. 336.
[162] LONDON GAZETTE. 16 May 1941. 2114 Honours and Awards Supplement. 2.

[163] LUFF. 2002. 331.

[164] Amy's mother Ciss died seventeen years later on 6 August 1958 and her father Will died five years after his wife (25 October 1963) at Woodmansey (a village to the south of Beverley on the road to Hull). It seems that he spent his final few years carefully tidying up crucial matters linked to his AJK fish business and Amy's artefacts. He liked to leave everything in order – it was in the nature of the man.

[165] ST.GEORGE'S WESLEYAN METHODIST MINUTE BOOK (1910-1951) from Chris Ketchell – see Endnote No.30 above.

[166] SMITH. 1967. 347.

[167] HARRISON, Amanda J. 2015. Website. Female Commercial Tiger Moth Pilot. http://amandajharrison.com/. Accessed 25 February 2015.
NOTE: When I doubled-checked her website in June there was no mention of Amanda's 'Australian Adventure'. I am not sure what the situation is since then – and I have never made direct contact.

[168] PETRE, Jonathan. 4 July 2015. Article. Amy flies again…repeated 85 years on by bold aviatrix? The Mail on Sunday. 1-8.
http://www.dailymail.co.uk/news/article-3149712/Amy-flies-solo-global-flight-tragic-death-inspired-world-Amy-Johnson-s-amazing-journey-repeated-85-years-bold-aviatrix-MoS-backing-just-that.html. Accessed 14 July 2015.
NOTE: (Spring 2016) Tracey Curtis-Taylor successfully completed her planned trip. She did well and I followed her progress via her FaceBook page. It was all good publicity and she honoured the memory of Amy Johnson along the way.

[169] WELTON, Rick. 2015. FaceBook Group. Amy Johnson Festival 2016. https://www.facebook.com/amyjohnson2016?pnref=story. Accessed 13 May 2015.

[170] PRISTON, Jane. 2015. Website. Amy Johnson & Herne Bay Project. http://www.amyjohnsonhernebay.com/. Accessed 27 January 2015.

[171] CRAWLEY. 1989. 26.

[172] The 'unbroken mirror' theory is interesting, but not conclusive proof that Amy landed her Ox-Box upon the Thames Estuary. Hull History Centre archivist Michéle Beadle – an authority on Amy's letters and life – has seen this mirror at Sewerby Hall. She does not wholly subscribe to this theory and stated that the mirror was still wrapped in its Christmas paper and might well have been given extra wrapping in clothing – as one does with a fragile gift in a light travel bag. (1:1 discussion 21 August 2015).

[173] WOMAN ENGINEER. 1941. 88. Cover letter to the article by Amy Johnson called A Day's Work in the ATA. 89-90. See Appendix 15F.

Made in the USA
Charleston, SC
29 July 2016